Activities Manual
for
LIFE IN THE UNIVERSE
Second Edition

Edward Prather
University of Arizona

Erika Offerdahl
University of Arizona

Timothy F. Slater
University of Arizona

San Francisco   Boston   New York
Capetown   Hong Kong   London   Madrid   Mexico City
Montreal   Munich   Paris   Singapore   Sydney   Tokyo   Toronto

Editorial Director: Adam Black
Senior Acquisitions Editor: Lothlórien Homet
Media Producer: Deb Greco
Assistant Editor: Ashley Taylor Anderson
Senior Marketing Manager: Scott Dustan
Managing Editor: Corinne Benson
Production Supervisor: Liz Winer
Production Service: Connie Strassburg/GGS Book Services
Main Text Cover Design: Marilyn Perry
Supplement Cover Design: Seventeenth Street Studios
Manufacturing Manager: Pam Augspurger

ISBN-10   0-8053-1712-0
ISBN-13   978-08053-1712-1

Copyright © 2007, 2003 Pearson Education publishing as Pearson Addison-Wesley, 1301 Sansome St., San Francisco, CA 94111. All rights reserved. Manufactured in the United States of America. This publication is protected by Copyright and permission should be obtained from the publisher prior to any prohibited reproduction, storage in a retrieval system, or transmission in any form or by any means, electronic, mechanical, photocopying, recording, or likewise. To obtain permission(s) to use material from this work, please submit a written request to Pearson Education, Inc., Permissions Department, 1900 E. Lake Ave., Glenview, IL 60025. For information regarding permissions, call (847) 486-2635.

Many of the designations used by manufacturers and sellers to distinguish their products are claimed as trademarks. Where those designations appear in this book, and the publisher was aware of a trademark claim, the designations have been printed in initial caps or all caps.

INTRODUCTION

The confirmed detection of extant life beyond our own planet holds the promise to impact all aspects of science, technology, and society as we know it. Astrobiology can be defined as the search for life in the universe. This search spans from understanding how microscopic bacteria are able to thrive in extreme environments here on Earth to searching for planets around distant stars. The topics addressed in astrobiology involve research from nearly all fields of science. As such, it is inherently an interdisciplinary science addressing such questions as (1) under what conditions does life arise and exist? and (2) where else in the universe might we find it?

Until recently, most of the scientific community had relegated the idea of life beyond Earth to the realm of science fiction. However, recent discoveries have dramatically changed the scientific community's views about the potential for life in the universe. To date, we have discovered more than ten times as many planets outside our solar system as there are within it, regardless of which definition of "planet" you choose to use. Perhaps even more impressive is that we have found life to not only survive, but to flourish on Earth under conditions previously thought impossible. These include organisms that thrive in temperatures above the boiling point and below the freezing point of water, in extremely acidic and basic conditions, thousands of feet below Earth's surface, on the ocean's floor, and in the adverse radiation conditions of outer space. These discoveries are being made concurrently with discoveries that strongly suggest that liquid oceans of water exist under the icy surface of Jupiter's moon, Europa, and that subsurface ice is likely present across vast regions of Mars.

The second edition of the *Life in the Universe Activities Manual* is designed with two main goals in mind. The first goal is to help students gain an in-depth understanding of the fundamental concepts that characterize astrobiology. The second goal is to engage students in developing critical reasoning skills. To accomplish this, we utilize a guided inquiry instructional approach that is distinctly different than traditional laboratory exercises.

ACKNOWLEDGMENTS

This second edition project is the culmination of years of collaborative work by numerous individuals and we would like to express our sincere appreciation to all of them, even though they are not all listed here. This includes generous support from Nick Wolff and Michael Meyer from the University of Arizona's LAPLACE Astrobiology Center, primarily funded by the NASA Astrobiology Institute. Initial funding for the project was provided by George Tuthill and Kim Obbink from the Montana State University NASA Center for Education Resources (CERES) Project from NASA HQ #NAG5-4576 who supported participating graduate students, teachers, and scientists who worked on various aspects of the project early on. These original participants include Mel Anderson, Buck Buchanan, Donna Governor, Janet Erickson, Josy Mclean, Linda Sauter, Sandra Shutey, Alison Simmerman, Chrystel Wells, and Marty Wells. We also benefited greatly from our colleagues Chris Impey, John Keller, Lloyd Magnuson, Michael Meyer, Dwight Taylor, Jason Peak, and our unknown external reviewers. We gratefully appreciate the undeterred and Herculean efforts of University of Arizona sabbatical visitors Tom Olien and Alex Storrs. However, in the end, it was the insightfulness and encouragement from Ashley Taylor Anderson, Jeff Bennett, Adam Black, Deb Greco, and Lothlorien Homet at Addison-Wesley who helped bring these materials successfully to publication and we are deeply grateful for their support.

E. E. Prather, E. G. Offerdahl, and T. F. Slater
Tucson, Arizona

Preface

NOTES FOR THE INSTRUCTOR

The activities in the second edition of the *Life in the Universe Activities Manual* are designed to be completed in 2 to 3 hours. In some cases, optional sections have been included to extend the activity if needed. It is important to note that these activities are structured so that they can be easily adapted to accommodate shorter instructional periods.

It is our experience that most students learn better through social interactions than they do by passively listening to lectures. As a result, we have designed these activities around a collaborative learning model in which students work in small learning groups of three to four members. We have found that having students work only in pairs or by themselves is often insufficient to promote an exchange of diverse student ideas necessary to effectively and accurately operationalize scientific concepts. Furthermore, groups larger than four inevitably allow some students to disengage and become observers.

Your role as the instructor is crucial to the success of these activities. As you circulate among student learning groups, you should carefully listen to students' discussions about the concepts and ensure that they negotiate their own ways of thinking about these ideas and formulate accurate answers in their own words. When students are struggling, resist the temptation to provide students with answers. Rather, ask questions that help students build from their current understanding and encourage group discussion to focus on the topic of investigation. Used in this manner, these activities can provide "teachable moments" and "aha's" in your classroom.

One of the activities requires special preparation by the instructor prior to its use. In Part A of *The Nature of Life* activity, you will need two types of plant specimens, one that is living and one that is not living. Examples are real carnations and silk carnations (available for purchase at most craft stores) for each student group. In Part B, each group will need three cups or Petri dishes, labeled A, B, and C. Each cup should contain approximately ½ cup dirt or sand and one tablespoon of sugar. In addition, cup B needs to contain a teaspoon of yeast and cup C needs to have one crushed Alka-Seltzer tablet. In preparing the three samples, be certain that there are no visually distinguishing differences between the cups. Part of the experiment involves the addition of water to the Petri dishes. Therefore it will be necessary to have an adequate supply of water on hand.

Preface

NOTE TO THE STUDENT

Fortunately, astrobiology is not about memorizing facts, figures, and formulas - how boring might that be! Instead, astrobiology focuses on investigating the nature of life on Earth and if life exists elsewhere in the universe. Astrobiology isn't just one half astronomy and one half biology. Rather, it is an interdisciplinary science that uses ideas from many different scientific fields including chemistry, engineering, geology, mathematics, meteorology, oceanography, philosophy, physics, and sociology. This makes the study of astrobiology exciting as it includes something for everyone's interests. If you find you don't particularly like the concepts being studied this week, hang on, next week will likely be very different.

This second edition of the *Life in the Universe Activities Manual* was designed carefully for you, the learner. You'll likely find these activities to be quite different from those in other classes. Rather than simply calculating a number from a formula or finding an example of a particular term, each numbered question asks you to carefully think about and apply a concept. Each activity is designed to be completed in small groups. Astrobiologists work in interdisciplinary science teams because no one knows everything nor has had the same experiences. It is crucial that you take the time to understand what each person in your group thinks and then, as a group, reach a consensus answer. Each group member must write a thorough response in his or her own words in their own individual copy of the activity. You'll want to write as detailed an answer as possible because it is likely you'll use these activities to study for upcoming exams.

Astrobiology is an exciting field and we are learning new things every day. As the field is growing at a phenomenal rate, your responsibility isn't to memorize everything. Rather, your role is to learn how scientific concepts can be used to explore the limitations on life as we know it and where else beyond Earth we might find it. Welcome to the study of life in the universe!

E. E. Prather, E. G. Offerdahl, and T. F. Slater
Tucson, Arizona

TABLE OF CONTENTS

ACTIVITY 1:	IS IT SCIENCE?	1
ACTIVITY 2:	THE UNIVERSE IS A REALLY BIG PLACE	9
ACTIVITY 3:	REMOTE SENSING: WHAT CAN WE SEE WHEN WE CAN'T TOUCH?	17
ACTIVITY 4:	THE EVOLVING EARTH: GEOLOGIC AND BIOLOGIC TIME	31
ACTIVITY 5:	THE NATURE OF LIFE	49
ACTIVITY 6:	DESIGNER GENES FOR A DESIGNER WORLD	55
ACTIVITY 7:	THE EXTREME ENVIRONMENTS OF EARTH AND THE CREATURES THAT LIVE THERE	69
ACTIVITY 8:	LIVING A POLAR LIFESTYLE: THE IMPORTANCE OF WATER FOR LIFE	85
ACTIVITY 9:	TO TERRAFORM OR NOT TO TERRAFORM MARS, THAT IS THE QUESTION	97
ACTIVITY 10:	INTERSTELLAR REAL ESTATE: DEFINING THE HABITABLE ZONE	111
ACTIVITY 11:	WOBBLING STARS: HOW EXTRASOLAR PLANETS ARE DISCOVERED	131
ACTIVITY 12:	THE RARE EARTH: HOW RARE IS EARTH-LIKE LIFE?	145
ACTIVITY 13:	THE DRAKE EQUATION: ESTIMATING THE NUMBER OF CIVILIZATIONS IN THE MILKY WAY GALAXY	157
ACTIVITY 14:	IS THERE ANYBODY OUT THERE?	165

1
IS IT SCIENCE?

GOALS
- Differentiate between science and pseudo-science
- Evaluate the sources of claims that sound scientific
- Apply skepticism to scenarios that claim to be scientific
- Evaluate data-based evidence in scientific claims

Science is a particular way of observing and making sense of the natural world. As a result, science tries to address only questions pertaining to the natural world that can be studied through systematic observation and experimentation. Science makes claims based on evidence that are testable or make testable predictions. In contrast, pseudo-science makes seemingly scientific claims that lack verifiable evidence and are inconsistent with the nature of data-based arguments which are commonplace in mainstream science.

PART A: TESTING THE PREDICTIONS OF ASTROLOGY USING THE RULES OF SCIENCE

Astrology is often described as the notion that the positions of stars and planets influence events and human affairs on Earth. It has ancient roots and is well-known today through a variety of modes, including newspaper and Internet media sources. Although somewhat of an oversimplification, your horoscope birth sign, or zodiac sign, corresponds to the constellation the Sun was covering at the moment of your birth. Therefore, your horoscope sign is said to influence your personality and your life pathway.

1. What is your birth date (month and day)?

2. From the table below, CIRCLE the horoscope sign that corresponds to your birthday.

Aries Mar 21 — Apr 19	*Leo* Jul 23 — Aug 22	*Sagittarius* Nov 22 — Dec 21
Taurus Apr 20 — May 20	*Virgo* Aug 23 — Sep 22	*Capricorn* Dec 22 — Jan 19
Gemini May 21 — Jun 21	*Libra* Sep 23 — Oct 22	*Aquarius* Jan 20 — Feb 18
Cancer Jun 22 — Jul 22	*Scorpio* Oct 23 — Nov 21	*Pisces* Feb 19 — Mar 20

3. The following page includes descriptions and characteristics associated with horoscope birth signs. Carefully consider all 12 and determine the description that seems to most closely match you.

Activity 1

INDIVIDUAL CHARACTERISTICS AS DETERMINED BY HOROSCOPE BIRTH SIGNS

1. LIKES: Optimistic and freedom-loving people; Good-humored jokes; Honesty; Intellectualism DISLIKES: Blindly optimistic people; Carelessness; Irresponsibility; Restlessness	**5.** LIKES: Speculative ventures Lavish living; Pageantry and grandeur ; Children; Drama DISLIKES: Doing things safely; Ordinary, day to day living; Small minded people; Penny pinching; Mean spiritedness	**9.** LIKES: Action; Coming in first; Challenges; Championing causes; Spontaneity DISLIKES: Waiting around; Admitting failure; No opposition; Tyranny; Other peoples advice
2. LIKES: Reliability; Professionalism; Knowing what you discuss; Firm foundations; Purpose DISLIKES: Wild Schemes; Fantasies; Go-nowhere jobs; Ignominy; Ridicule	**6.** LIKES: Health foods; Lists; Hygiene; Order; Wholesomeness DISLIKES: Hazards to health; Anything sordid; Sloppy workers; Squalor; Being uncertain	**10.** LIKES: Stability; Being attracted; Things natural; Time to ponder; Comfort and pleasure DISLIKES: Disruption; Being pushed too hard; Synthetic or "human made" things; Being rushed; Being indoors
3. LIKES: Fighting for causes; Dreaming and planning for the future; Thinking of the past; Good companions; Having fun DISLIKES: Full of air promises; Excessive loneliness; The ordinary; Imitations; Idealistic	**7.** LIKES: The finer things in life; Sharing; Conviviality; Gentleness; Diplomacy DISLIKES: Violence; Injustice; Brutishness; Being a slave to fashion	**11.** LIKES: Talking; Novelty and the unusual; Variety in life; Multiple projects all going at once; Reading DISLIKES: Feeling tied down; Learning, such as school; Being in a rut; Mental inaction; Being alone
4. LIKES: Solitude to dream in; Mystery in all its guises; Anything discarded to stay discarded; The ridiculous; Likes to get 'lost' DISLIKES: The obvious; Being Criticized; Feeling all at sea about something; Know-it-alls; Pedantry	**8.** LIKES: Truth; Hidden Causes; Being involved; Work That is Meaningful; Being Persuasive DISLIKES: Being given only surface data; Taken advantage of; Demeaning jobs; Shallow relationships; Flattery and flattering	**12.** LIKES: Hobbies; Romance; Children; Home and country; Parties DISLIKES: Aggravating situations; Failure; Opposition; Being told what to do; Advice (good or bad)

Adapted from http://www.astrology-online.com

Is It Science?

4. One might argue that individuals who are able to rise to high public office, the president of the United States for example, might exhibit similar traits. If true, which numbered horoscope description (or descriptions) do you think most closely matches this type of individual?

5. Now imagine that there is no direct correlation between horoscope sign and individuals who rise to high public office. How would you expect the sign(s) of these individuals to be distributed among the 12 descriptions provided?

The last few pages of this book provide a table showing which descriptions match which horoscope signs.

6. Did the description you selected correspond to your horoscope sign?

7. How many of your classmates sitting near you correctly selected their horoscope sign?

8. What percentage of students sitting near you correctly selected their horoscope birth sign?

$$\frac{\text{Number of students identifying correctly}}{\text{Number of students in the class you asked}} \times 100 = _____\%$$

9. Does your answer to Question 8 support the hypothesis that astrology works (i.e., that birth signs can be used to predict events and personality traits) or does it demonstrate that astrology does not work (i.e., that people do not identify their own birth signs more often than would be expected by random guessing)? Justify your answer.

10. As it turns out, the birthdays of U.S. presidents are evenly spread out across all months. Does this confirm the idea in Question 4 or in Question 5? What does this imply about the scientific validity of astrology?

Activity 1

PART B: EVALUATING SCIENTIFIC SOURCES

Claims that sound scientific come to us from a variety of media sources. Imagine you saw the following two stories.

News Item A	News Item B
Headline: *Cancer Risk Reductions May Lie in Foods We Eat*	Headline: *Blazing Laser Gun Cures Cancer from 1 Mile Away*
Publication: *New England Journal of Medicine*	Publication: *Healthy Living Times*
Author(s): *Susan Yi and D. L. Strendinsky, Davidson-Smith University*	Author(s): *J. Clarke, Center for Healing without Narcotics — Division of Laser Sales and Marketing*

1. When evaluating the scientific credibility of a news item, the first step is to consider if the claim is plausible from a scientific perspective. Which claim do you think seems more plausible, A or B? Why?

2. Which publication do you think seems most scientifically credible, A or B? Why?

3. Which author do you think seems most scientifically credible, A or B? Why?

4. When evaluating the validity of a scientific-sounding claim, you need to consider the plausibility of the claim, the credibility of the source, its authors, and if the authors could have any ulterior motivation for spreading the idea. Consider the following comment from a student about information found on the Internet.

   I was reading an EndofHumanity.com *Web page a diagram showing that the both Uranus and Neptune will be on the same side of the Sun as Earth on June 26, 2016 and that terrible earthquakes will result causing the end of human existence on Earth. They suggested that we donate 10% of our life savings to their Web site so that they can more widely disseminate this crucial information. I think it is critical that we all participate in this important cause because it is really going to happen!*

   a. Why would one be suspect of the claim's scientific plausibility?

   b. Why would one be suspect of credibility of the claim's source or authors?

   c. Why would one be suspect of a possible ulterior motive?

Here is a small segment from a recent news story. Do you believe it? In the space below, what are the characteristics that make it believable what make it unbelievable?

SCIENTISTS SUCCESSFULLY BREED 7-LEGGED SPIDER

<Reuters News Service> A spokesperson for The Lamarkian Augmentation Company (LA Co.) announced at a press conference that LA Co. scientists had successfully bred spiders with seven legs. Speaking on behalf of the six scientists present, Dr. John Zoriski, who recently left his post as a professor at Florida Institute of Technology and Agriculture, stated, "Breeding a 7-legged spider is relatively easy. All you have to do is mate one male spider and one female spider that have each had one of their eight legs surgically removed and their offspring will only have seven legs." The team reported that they had an 82.71% success rate since their first experiment in 2005 with the 8-legged species *Araneae actinopodidae*, but waited to announce the results until they had achieved an acceptable success rate with another 8-legged species, *Poecilotheria regalis*, the standard model species used by the US National Institute of Health. Dr. Patricia Robinez, an expert in molecular biology techniques independently underscored the importance of these results when she exclaimed, "Wow, this is a fantastic claim!" upon hearing the announcement. Dr. Zoriski's team's results will be submitted next month to the scientific journal, NATURE, one of the world's most well-recognized scientific journals.

5. Complete the table below, based on the answers you gave on the previous page.

DO YOU BELIEVE IT?	
Characteristics That Make This Story Believable	*Characteristics That Make This Story Unbelievable*
i.	i.
ii.	ii.
iii.	iii.
other reasons:	other reasons:

6. What additional evidence could the authors have provided that you would accept as convincing for the story's claim about the ability to breed seven-legged spiders? Explain your reasoning.

Activity 1

PART C: SCIENTIFIC EVIDENCE

Scientists make several common assumptions about the world. Some of these important assumptions described in 1990 by AAAS book, *Science for All Americans*, can be summarized as:

- Science explains and predicts
- Scientific knowledge can change
- Science cannot provide complete answers to all questions
- Scientific conclusions require reproducible evidence
- Science requires a blend of logic and imagination
- Scientists share a worldview that the universe is understandable
- Scientists try to identify and avoid bias
- Science is not authoritarian

1. Consider the scientific-sounding claim shown in the box below and possible list of evidence in support of that claim. Based on the assumptions listed above, which single line of evidence (A, B, C, or D) would you accept as being most convincing (*circle one*)?

CLAIM	POSSIBLE LINES OF EVIDENCE (Circle the Most Convincing)
Toilets do NOT always flush clockwise in the Northern Hemisphere and counterclockwise in the Southern Hemisphere, even though the Coriolis effect exists.	A. Because hurricanes spin in opposite directions in the Northern and Southern Hemispheres, one can infer that this is true for toilet flushings as well. B. The spinning direction of 100 toilets in the Northern Hemisphere city of New York City and 100 toilets in the Southern Hemisphere city of Sidney are recorded and compared. C. The spinning direction of 10 toilets in the Northern Hemisphere cities spread across the globe (including Los Angeles, Chicago, London, Paris, and Moscow) are recorded and compared. D. A toilet is flushed in Miami on four different dates-once each when the moon was new, at first quarter, full, and at third quarter- and the direction of spinning was noted.

Explain why you circled the line of evidence you did.

2. Consider the scientific-sounding claim shown in the box below. Describe the scientific evidence that would overwhelmingly compel you to believe this claim scientifically.

CLAIM	COMPELLING EVIDENCE
Smoking causes lung cancer	

3. Consider the scientific-sounding claim shown in the box below. Create and describe the specific steps in an experiment that you would conduct or what evidence you would need to collect to fully confirm this claim scientifically.

CLAIM	EXPERIMENT TO TEST THIS CLAIM
Tapping the side of a soda can will prevent its contents from foaming over when you open it.	

Explain why you chose that experiment or evidence.

Activity 1

4. Invent a scientific sounding claim, much like those listed above, that is NOT actually scientific and explain why it is not scientific.

5. Now, rewrite your scientific-sounding claim to be scientifically credible and explain specifically why it is now scientific.

2
THE UNIVERSE IS A REALLY BIG PLACE

GOALS
- Gain an intuitive feel for different measures of distance
- Convert between various units of distance
- Use scientific notation
- Understand concepts of the astronomical unit and light-year
- Create scale models of solar system and galaxy using a campus map

As we begin our study of life in the universe, you'll find that our universe is indeed a really big place. The distances between planets are unimaginably large, and the distance to the nearest star system is even larger. The activities that follow ask you to carefully create several scale models and use them to reason about the size and scale of the universe.

PART A: CONVERTING BETWEEN UNITS

1. Name a location that is about a 1-hour drive from your present location.

2. Approximately how far away (in miles) is the location you listed above?

Instead of miles, astrobiologists use a standard of measure that is used internationally, the kilometer (km). Therefore, it will be useful for you to begin to think of distances in these terms.

Using your calculator, you can convert between miles and kilometers using the relationship that 1.609 kilometers = 1 mile. For example:

$$15.0 \text{ miles} \times \frac{1.609 \text{ km}}{1 \text{ mile}} = 24.1 \text{ km or about 24 km}$$

Note that we keep only three "significant" figures in the final answer. This is because there were only three figures in the initial value 15.0 miles.

3. Calculate the number of kilometers for the distance you estimated in Question 2. Clearly show all of your work as in the example calculation above.

4. Name a location that would take about 3 hours to drive to and estimate the number of kilometers to that location.

5. Current estimates suggest that a liquid ocean may exist about 100 km beneath the icy surface of Jupiter's moon Europa. Give examples of two locations on Earth that are about 100 km apart.

Activity 2

When dealing with very large numbers, you can easily enter them into your calculator using scientific notation. Scientific notation is a kind of shorthand for large numbers used by scientists. There is a special button on most scientific calculators that does this for you. For example, 93,000,000 is typically written in scientific notation as 93×10^6 and can easily be entered into your calculator as 93 EXP 6 or possibly as 93 EE 6 or 93 EE X 6. In other words, the EXP or similar calculator key does the *X 10* operation for you.

6. For distances as large as the distance between planets, scientists frequently use a unit called the *astronomical unit*, or AU. An AU is the average distance between Earth and the Sun (about 93 million miles). Calculate how many million kilometers there are in an AU.

7. If Jupiter is five times farther from the Sun than Earth is, how many AU are between Jupiter and the Sun?

8. Venus orbits the Sun at a distance of 0.7 AU. Mars orbits the Sun at a distance of 1.5 AU. Provide a sketch of the orbits of Venus and Mars that shows the relative distances of these planets' orbits around the Sun.

9. When Venus and Mars are closest to one another during their orbits around the Sun, how far apart are they?

10. When Venus and Mars are farthest from one another during their orbits around the Sun, how far apart are they?

PART B: SCALE OF THE SOLAR SYSTEM

To better appreciate the unimaginable distances between objects in our solar system, it is useful to construct a scale model that depicts these distances in a more familiar context. We will use *scale factors* to convert the actual distances between objects within the solar system to a distance that appropriately fits our scale model. A scale factor is simply "the desired size you would like for your model" divided by the "actual size" of the thing being modeled.

$$\text{scale factor} = \frac{\text{desired size}}{\text{actual size}}$$

In this activity we will construct a scale model of our solar system to fit on a 100-yard football field. The size of our solar system is roughly equivalent to the average distance between the Sun and Pluto, about 40 AU. The scale factor for our football field-sized scale model will be:

$$\text{scale factor} = \frac{\text{(desired size of scale model)}}{\text{(actual distance between Sun \& Pluto)}} = \frac{100 \, yards}{40 \, AU}$$

$$\text{scale factor} = \frac{100 \, yards}{40 \, AU}$$

scale factor = 2.5 yards per AU or 2.5 $^{yards}/_{AU}$

Then you will calculate the distance between the Sun and each of the planets using your scale factor. For example, Mercury is 0.4 AU from the Sun. Therefore, the distance between the Sun and Mercury on the football field would be:

$$\text{Distance between Sun and Mercury} = 0.4 \, AU \times 2.5 \frac{yards}{AU} = 1 \, yard$$

1. Determine an appropriate scale factor for a scale model of the solar system that will fit on to a football field and record it below. Use the scale factor of 2.5 $^{yards}/_{AU}$. The first one has been filled in for you.

PLANET	APPROXIMATE DISTANCE FROM SUN (IN AU)	DISTANCE FROM SUN (LOCATED ON THE GOAL LINE) IN YARDS
Mercury	0.4	0.4 AU x 2.5 $^{yards}/_{AU}$ = 1 yard
Venus	0.7	
Earth	1.0	
Mars	1.5	
Jupiter	5.2	
Saturn	9.5	
Uranus	19.2	
Neptune	30.1	
Pluto*	39.5	

Pluto may or may not formally be defined as a planet at this moment.

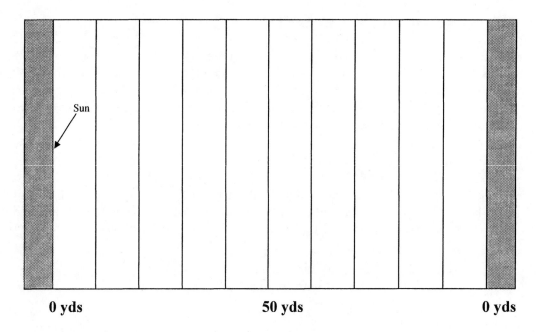

0 yds   **50 yds**   **0 yds**

2. Place labels clearly showing where each planet should be placed on the sketch of a football field above.

3. Although Mars is farther from the Sun than Earth, Mars is often called an inner planet. Why do you think Mars is called an inner planet?

4. How do the spacings of the inner planets compare to the spacings of the outer planets?

5. The farthest astronauts have traveled is to the Moon, which is located about 0.003 AU from Earth and represents a 3-day journey one-way. Calculate the number of yards (using your scale factor) between Earth and the Moon.

6. Is it possible to show the Moon's orbit around Earth on your football field scale model diagram? Why or why not?

7. If it takes 8 years for a spacecraft to travel from Earth to Jupiter, estimate the minimum time it would take to travel from Earth to Saturn. Explain how you arrived at your estimate.

PART C: PROPOSING A SCALE MODEL OF THE SOLAR SYSTEM ON YOUR CAMPUS

In this activity, your group will develop a proposal for a scale model of the solar system on your campus (Note: You may wish to choose an alternative nearby location such as a shopping mall or downtown area rather than your campus for your scale model if desired.)

1. Use a copy of a campus map or a rough sketch on a separate piece of paper to locate distinguishable campus landmarks. Pick two campus locations to place your scale model between. Clearly indicate where you want to place the Sun and Pluto.

2. What is the distance between the Sun and Pluto on your map or sketch?

3. Using the distance in Question 2 as the desired size of your scale model, calculate the scale factor (as you did in Part B) that you will use for your campus scale model of the solar system.

4. Complete the table below using your scale factor. Then clearly label the approximate position of the Sun and the planets on your map or sketch.

PLANET	APPROXIMATE DISTANCE FROM SUN (IN AU)	APPROXIMATE PLACEMENT ON SKETCH (NEAR WHAT OBJECT) AND DISTANCE FROM SUN IN APPROPRIATELY SCALED UNITS
Mercury	0.4	
Venus	0.7	
Earth	1.0	
Mars	1.5	
Jupiter	5.2	
Saturn	9.5	
Uranus	19.2	
Neptune	30.1	
Pluto*	39.5	

Pluto may or may not formally be defined as a planet at this moment.

Activity 2

5. Determine the appropriate distances to each of the following objects on your scale model and label them on your map or sketch (if possible):

   a. Moon's orbit at 0.003 AU from Earth

   b. Asteroid Ceres 2.8 AU from the Sun

   c. Comet Halley at 18 AU from the Sun

   d. *Voyager* space probe at 84 AU from the Sun

6. Alpha Centauri is the closest star system to the Sun at 272,000 (2.72×10^5) AU.

   a. Calculate the distance to Alpha Centauri for your campus scale model. Show your work.

   b. If you were to put this on your scale model, would it be on your campus, within your city limits, in your state, on your continent, or anywhere on Earth?

7. The Sun is located 2×10^9 AU from the center of the Milky Way Galaxy.

   a. Calculate the distance to the center of the Milky Way Galaxy for your campus scale model. Show your work.

   b. If you were to put this on your scale model, would it be on your campus, within your city limits, in your state, on your continent, or anywhere on Earth?

8. The Andromeda Galaxy is the largest galaxy closest to our own at 164×10^9 AU.

   a. Calculate the distance to the Andromeda Galaxy for your campus scale model. Show your work.

   b. If you were to put this on your scale model, would it be on your campus, within your city limits, in your state, on your continent, or anywhere on Earth?

PART D: SCALE OF THE UNIVERSE

In everyday language, we often use time as a way to describe distance. For example, one might say that it is about a 6-hour flight from San Francisco to New York City or that it is a 3-hour train ride from New York to Washington, D.C.

1. In units of time, estimate how long it is from where you are right now to the nearest state capitol using:

    a. a bicycle

    b. a car

    c. an airplane

For distances between different stars (interstellar distances) and between large collections of stars called galaxies (intergalactic distances), scientists use a standard unit of distance called a *light-year*, which is the distance light travels in one year. One light-year, abbreviated *ly*, is about 63,216 AU.

2. Imagine you are driving down a highway and see a billboard advertisement that reads, "This new computer technology is light-years ahead of its time." What is incorrect about the usage of *light-years* in this sentence?

3. Calculate how many light-years away Alpha Centauri is from Earth. (Hint: Consider how many AU Alpha Centauri is from Earth, Part C, Question 6.)

4. If you could travel in a spaceship at the speed of light, 300,000 kilometers per second (km/sec), approximately how many years would it take you to travel from Earth to the next star system, Alpha Centauri? (You don't really need a calculator for this calculation.)

5. Would the time you estimated in Question 3 be significantly different if you traveled from the Sun to Alpha Centauri instead of from Earth? Explain your reasoning.

6. Our home galaxy, the Milky Way Galaxy, has a diameter of about 100,000 light-years. If Earth is about halfway out from the center of our Milky Way, about how many light-years is Earth from the center?

Activity 2

7. The Andromeda Galaxy is located about 2.6 million light-years away from the Milky Way. How many times greater is the distance between the Milky Way and Andromeda compared to the distance between Earth and the center of the Milky Way?

8. The most distant galaxies known are about 10 billion light-years away. How many times farther away is the most distant galaxy compared to the distance to the Andromeda Galaxy?

9. Put the following objects in order from *closest* to Earth to *farthest* from Earth.

   a. Andromeda Galaxy
   b. The Sun
   c. Most distant galaxies
   d. Alpha Centauri
   e. Pluto
   f. The center of the Milky Way Galaxy

3
REMOTE SENSING: WHAT CAN WE SEE WHEN WE CAN'T TOUCH?

GOALS
- Explore the idea of spectral lines as signatures for the presence of different substances
- Interpret data from remote sensing images
- Understand how to use the data from remote sensing images to determine the probability of life beyond Earth
- Become familiar with the range of wavelengths in the electromagnetic spectrum

Astrobiology often involves the study of objects that cannot be touched or even seen directly. To do this we typically look for clues that indicate the presence or behavior of objects that cannot be directly observed. When scientists want to study such objects, they often make observations using a technique called *remote sensing*. With remote sensing, information about an object is gathered from a distance without ever touching or possibly even directly seeing the object with the naked eye.

PART A: SEEING THE LIGHT?

In Figure 1 below a satellite is used to gather images of Earth's surface. This satellite, like your eye, is sensitive only to visible light. Although the Sun gives off many other types of light, in this part of the activity we will be concerned with only the visible light (including all the colors of the rainbow) given off by the Sun.

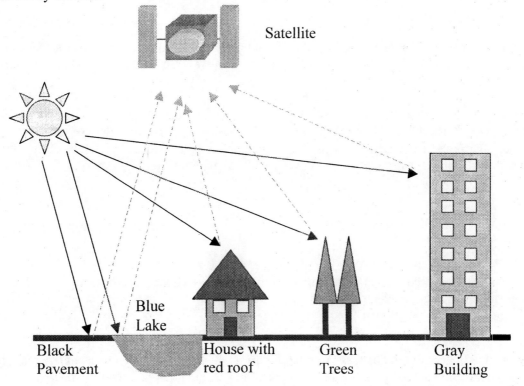

Figure 1

Life in the Universe – Activities Manual, 2nd Edition

Activity 3

1. Would the different objects all look the same to the satellite? What would appear different about the objects?

2. What color (or colors) will the satellite detect from the trees?

3. Which color (or colors) of light reaches the trees from the Sun?

4. Which color (or colors) of light is sent from the trees to the satellite? Explain how you know.

5. Where do the remaining colors of light that were sent by the Sun go once they reach the trees? Explain your reasoning.

6. Would the satellite still detect the roof of the house to be red late at night, long after the Sun had set? Explain your reasoning.

In the previous questions, we investigated how one can use a satellite to perform remote sensing and determine the color of objects on the surface of a planet. However, remote sensing can also be used to investigate the temperature of objects.

7. Imagine that your roommate bought two identical potatoes at the store. She put one potato in a cold freezer (0^oC) and the other potato in a warm oven (40^oC). Sometime later she took the potatoes out of the freezer and oven and placed them on the table. Could you tell which potato is cold and which one is warm just by looking at them? Explain why or why not.

Now imagine that you could not get close enough to the potatoes to hold them, but still wanted to determine which was hot and which was cold. This is the type of problem that is encountered when one does remote sensing.

Consider the two images of the potatoes shown in Figure 2. Image 1 was taken by a camera that, like our eye, is sensitive to visible light. Image 2 was taken by a camera that detects **only** *infrared light* (a form of light that is not visible to the naked eye). Scientists often look at the infrared light emitted by an object because this form of light can be used to determine the temperature of many common objects on Earth. In a typical infrared image, objects that appear bright are at higher temperatures and objects that appear dark are at colder temperatures.

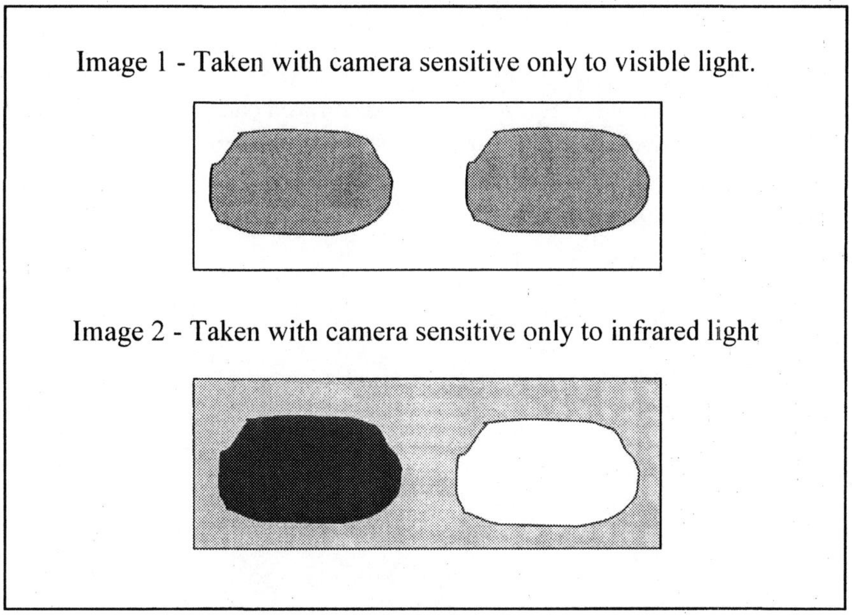

Image 1 - Taken with camera sensitive only to visible light.

Image 2 - Taken with camera sensitive only to infrared light

Figure 2

8. Which potato is hot? Which potato is cold? Which type of light (visible or infrared) did you use to determine your answer? Explain your reasoning.

9. Describe a situation in which both of the potatoes would be invisible to the human eye but not to an infrared camera.

10. Draw a sketch of what the infrared picture would look like the next day after the potatoes were left on the table overnight.

Activity 3

11. Draw an overhead sketch (looking straight down) of what the satellite in Figure 1 would detect during a hot summer day if it were equipped with an infrared camera. Be sure to properly shade your drawing to illustrate the differences in temperature of the objects.

PART B: IDENTIFYING THE LINES OF LIFE

We have seen that when light interacts with an object some of the light is absorbed and some of the light is reflected or re-emitted by the object. By examining how different wavelengths of light are absorbed and reflected (or re-emitted) we can determine information about the composition of the atmosphere or surface of an unknown planet or moon. In the graph shown below in Figure 3 (called an *absorption spectrum*), the percentage of light absorbed by water in Earth's atmosphere (as measured between 0.7 μm and 2.2 μm) has been plotted over a range of wavelengths. Recall that a micrometer (μm) is equal to 10^{-6} meters (m).

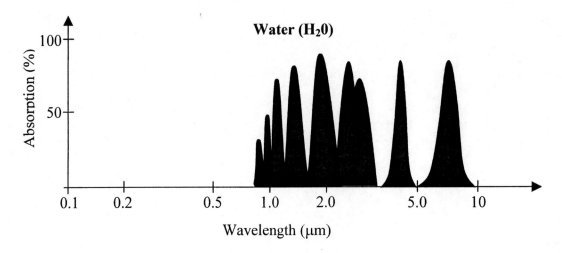

Figure 3

In the table below we have identified the approximate range of wavelengths and corresponding names (types) for the entire spectrum of electromagnetic radiation (light).

TYPE OF ELECTROMAGNETIC RADIATION (LIGHT)	WAVELENGTH (METERS)
Gamma ray	$< 10^{-11}$ m
X-ray	10^{-11} m $- 10^{-8}$ m
Ultraviolet	10^{-8} m $- 4 \times 10^{-7}$ m
Visible (VIBGYOR)	4×10^{-7} m $- 7 \times 10^{-7}$ m
Infrared	7×10^{-7} m $- 10^{-4}$ m
Microwaves	10^{-4} m $- 10^{-2}$ m
Radio	10^{-2} m $- > 10^{2}$ m

1. Identify the range of wavelengths of light that were observed to make the absorption spectrum for water shown in the graph in Figure 3. What is the name (or names) of the type of light that was used to collect this data?

2. Using the graph in Figure 3, list the value for each peak wavelength of light that is strongly absorbed by water.

The specific wavelengths of light that are absorbed or reflected from an object together make up a signature (like a fingerprint) that can be used to identify the specific composition of the observed object. Furthermore, there are specific conditions and indicators that are used by scientists to identify the presence of life. For instance, life as we know it on Earth requires the presence of liquid water, and the majority of liquid water here on Earth is found between the temperatures of 0°C and 100°C. Therefore, in our search for life in the solar system, it is essential to look for signs that liquid water is present or has been present in the past. Scientists have also identified other chemical signatures that can be used to indicate the presence of life. For instance, we might look for oxygen (O_2), carbon dioxide (CO_2), or methane (CH_4) that has been given off by living organisms. At the end of this activity we have provided a *Spectra Catalog* that identifies a set of key wavelengths that correspond to many of the molecules that are central to the search for life on Earth, in our solar system, and beyond our solar system.

Activity 3

3. What are the wavelengths that are identified in the *Spectra Catalog* for water? How do these wavelengths compare to the wavelengths that you identified in Question 2?

4. Over what range of wavelengths would you need to observe if you were planning on identifying all the molecules listed in the *Spectra Catalog* in your search for life?

5. On the axes provided below, sketch what the absorption spectrum for each of the molecules presented in the *Spectra Catalog* might look like. Pay less attention to the width and height of the peaks that you draw, and concentrate on drawing the peaks at the correct wavelengths.

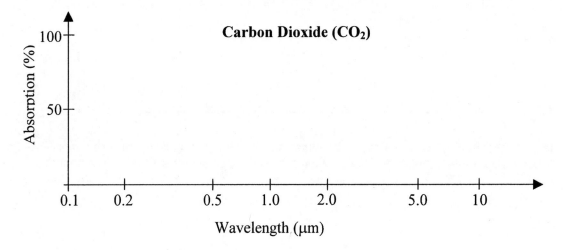

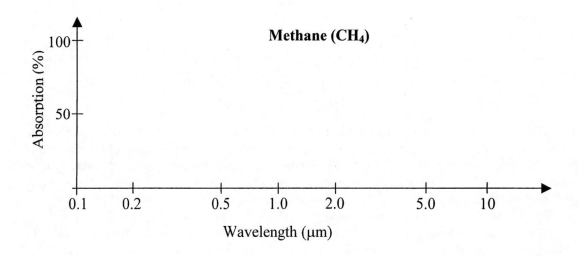

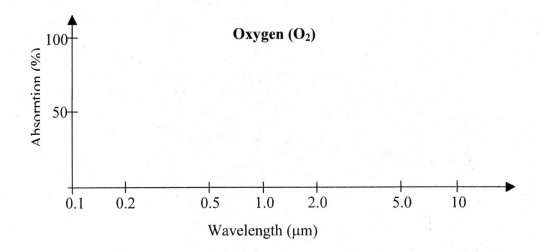

PART C: LOOKING AT THE DISTANT EARTH

To begin our search for life using remote sensing, we will consider what it would be like if our home were a distant planet from Earth and we had sent out a probe from our home planet that had reached the "mysterious" Earth.

1. What type of light would your probe need to be able to detect in order to search for life on Earth? Explain your reasoning.

Imagine that the probe is in an orbit directly above a region of intense biological and geothermal activity in Yellowstone National Park. The detector on your probe is able to gather data for the presence of different molecules near (or at) the surface of a planet.

2. During your survey of this region you determine that water (H_2O), methane (CH_4), and carbon dioxide (CO_2) are all abundant. List the wavelengths that may have been detected by your probe that would allow you to identify the presence of these molecules.

3. Identify the organism(s) or biological process(es) that could have produced the molecules identified in Question 2.

Activity 3

Now imagine that your probe took the grayscale image shown below in Figure 4 of the Old Faithful geyser with an infrared (IR) camera. It is found online at
http://coolcosmos.ipac.caltech.edu/image_galleries/ir_yellowstone/oldfaith1_gallery.html .

Figure 4

4. Which of the shaded regions of this image do you think correspond most directly to the location of the geyser? Explain your reasoning.

5. The hottest temperatures represented in this image are 100°C (and above) and the coolest temperatures are near 10°C. Which shades most likely represent higher temperatures and which shades most likely represent cooler temperatures?

6. Describe how this image and the data collected from Question 2 could be used to support the argument that life exists on the distant and mysterious Earth.

Part D: Do We See the Signs of Life in Our Solar System?

In this part of the activity will examine the other planets or moons in our solar system and consider what information we might be able to gather about these objects through remote sensing. In each case, you should consider whether the information provided suggests that life is or is not likely to occur on the planet or moon being investigated.

Imagine that your probe has left its orbit above Earth and has established a new orbit above the planet Mars. The grayscale image shown in Figure 5 below provides two pieces of information. On the left is an image of Mars in the visible part of the spectrum. The right image was taken of the surface of Mars with an infrared camera. The different shades of gray in the IR image are used to indicate the range of temperatures (from -120°C to 0°C) observed across the Martian surface. Note the dark shade of gray indicating -120°C that has been provided in the temperature scale to the right of the IR image. It comes from the THEMIS Public Data Releases, Mars Space Flight Facility, Arizona State University. (September 26, 2006) <http://themis-data.asu.edu>.

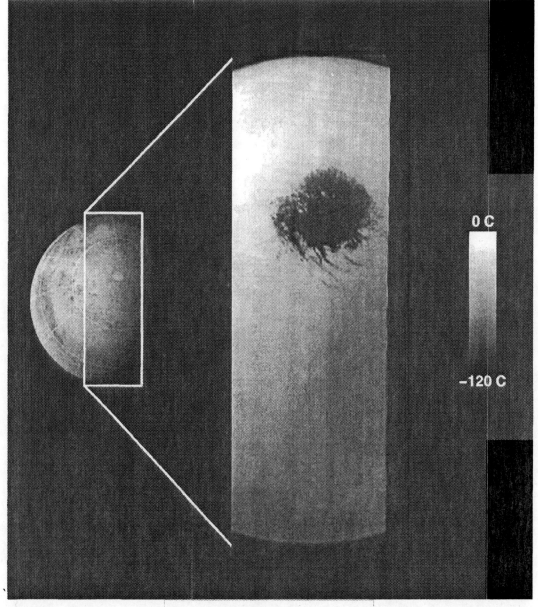

Figure 5

Activity 3

1. In this grayscale image of the planet Mars shown in Figure 5, what type of information is being displayed?

2. What is the range of values of the data?

3. Shade in the rest of the temperature scale with the approximate shades of gray to correspond with the information provided in the IR image of Mars.

4. In the IR image of Mars there are two shaded regions that each correspond to -120°C. One of these regions corresponds with the polar ice cap. Why do you think the other region of the Martian surface is as cold? (Hint: Examine the image of Mars on the left taken in the visible part of the spectrum. Consider how the light from the Sun is illuminating the surface of Mars.)

MARS INFORMATION

Atmosphere	Very thin; consists of 95% carbon dioxide, 3% nitrogen, 1.5% argon, and trace amounts of water.
Average Surface Temperature	-53°C (-63.4°F) Temperature occasionally rises above the freezing point in equatorial regions, although the pressure is too low for liquid water to exist on the surface.
Surface Info	Current evidence points to subsurface water (dirty ice) at latitudes at south and north between 60° and 90°.
Polar Ice Caps	Consist of CO_2 ice, water ice, and dust.

5. If your probe was to analyze the Martian atmosphere, which wavelength(s) (from the *Spectra Catalog*) would dominate the absorption spectra from the data you collect? Explain your reasoning.

6. Does the information presented in the IR image, or in the table, suggest that liquid water could be located on the surface of Mars? If so, in what region(s) of the planet is it most likely to be found? If not, why not?

7. Based on the data presented in the IR image and in the table, do you think that it is likely that life exists on Mars? Explain your reasoning.

The table below provides data on Europa (a moon of Jupiter).

EUROPA INFORMATION

Atmosphere	Extremely thin; major constituent is oxygen.
Average Surface Temperature	-150°C (-238°F)
Surface Info	Primarily water ice with strong evidence of water oceans beneath thick water ice crust. Water ice crust may be subject to partial melting due to tidal heating.

8. If your probe was in orbit above Europa, which wavelengths (from the *Spectra Catalog*) would dominate the absorption spectra from the data you collect?

9. Based on the information provided in the table, is it likely that there is life on Europa? Explain why or why not.

The table below provides data on Titan (a moon of Saturn).

TITAN INFORMATION

Atmosphere	Very thick; consists of 90% nitrogen and smaller amounts of methane, ethane, and argon.
Average Surface Temperature	-180°C (-292°F)
Surface Info	Composed of rock and ice (ice is primarily a combination of water, methane, and ammonia), oceans that are likely a mixture of ethane and methane, and possibly deep underground oceans of ammonia and water.

Activity 3

10. If your probe was in orbit above Titan, which if any of the wavelengths (from the *Spectral Catalog*) would be present in the absorption spectra from the data you collect?

11. Based on the provided information in the table, is it likely that there is life on Titan? Explain why or why not.

12. In Parts C and D you considered data collected by a probe that was sent to Earth, Mars, Europa, and Titan. Based on the remote sensing information gathered from these different locations, which planet or moon would you first select for a mission to further search for life? Which location would be your second choice? Third? Fourth? Be sure to include any necessary scientific findings that the initial probe mission found that influenced your ranking.

Spectra Catalog

Oxygen (O_2)

- Spectrum peak wavelengths at 0.69 μm, 0.760 μm, and 1.28 μm.

- Oxygen is a by-product of oxygenic photosynthesis, a metabolic process performed by plants as well as certain types of bacteria, such as the bacterium *Cyanobacteria*. It is also produced by the photodissociation of water.

Carbon Dioxide (CO_2)

- Spectrum peak wavelengths at 2.0 μm, 2.06 μm, 2.7 μm, 4.3 μm, and 15 μm.

- Carbon dioxide is a by-product of several biological processes including respiration, a metabolic process performed by many organisms. These organisms include animals and many types of bacteria, such as the *Sulfolobus* bacteria. *Sulfolobus* are found in Yellowstone National Park at temperatures up to 90°C.

Methane (CH_4)

- Spectrum peak wavelengths at 3.3 μm and 7.65 μm.

- Methane is another by-product of metabolism. An example of a living organism that produces methane is the type of bacteria called *Methanopyrus* (capable of growth in temperatures of up to 110°C). Methane is also formed in planetary atmospheres as a result of photochemistry. Methane can be converted chemically into many other organic compounds.

Water (H_2O)

- Spectrum peak wavelengths at 0.820 μm, 0.950 μm, 1.13 μm, 1.38 μm, and 1.87 μm.

- Water can be found as a gas, solid, or liquid. It can be photodissociated into hydrogen and oxygen. In its liquid form, it is a solvent for life. Water is the by-product of several biologic processes, including respiration.

4
THE EVOLVING EARTH:
GEOLOGIC AND BIOLOGIC TIME

GOALS
- Examine the geological and biological events of Earth's past
- Explore the causal relationship between Earth's geology and biology
- Create relevant time scales

PART A: ESTABLISHING A TIMELINE

As a group, use the set of six *Timeline Cards* found at the end of the activity to complete the following questions.

1. Using the information provided on each card, list the cards in the table below in *chronological* order from the earliest events on Earth to the most recent events on Earth. Next to the name of each time period, state how many years it lasted.

	TIME PERIOD	DURATION (YRS)
1		
2		
3		
4		
5		
6		

2. Now create a timeline on a single sheet of paper turned sideway in front of you. Using a ruler, draw a horizontal line across the middle of the paper so that it runs almost the entire length of the paper. Add tick marks to your line to indicate the beginning and ending of each of the six time periods. The distance between the beginning and ending marks of a time period should represent the relative amount of years elapsed during each period.

3. Are the time intervals for all the time periods equal? Which is longest? Which is shortest?

4. In which era or eon was Earth 3.9 billion years ago?

Activity 4

PART B: GEOLOGY THROUGH THE AGES

Now refer to the set of *Geology Cards* at the end of the activity. Carefully read the information on each of the six cards. By discussing the events listed on the cards, decide as a group how the cards should be placed for them to be in *chronological* order (from the earliest events on Earth to the most recent events).

1. In what order did you place the *Geology Cards*?

	GEOLOGY CARD
1	
2	
3	
4	
5	
6	

2. As Earth evolved from one time period to the next, identify one critical change or event (piece of geological evidence) that enabled your group to order the cards in the way you did.

TIME PERIOD TRANSITION	CRITICAL CHANGE OR EVENT
Hadean Eon → Archaean Eon	
Archaean Eon → Proterozoic Eon	
Proterozoic Eon → Paleozoic Era	
Paleozoic Era → Mesozoic Era	
Mesozoic Era → Cenozoic Era	

Evolving Earth

Match each of your *Geology Cards* with the corresponding time period from your *Timeline Cards*. By placing the cards in order, you have created a geological timeline of events in Earth's past.

Get together with another group. Was your timeline the same as the group with which you compared your results? If not, how would you change your group's and/or the other group's timeline so they match? Describe the geological evidence that influenced any changes.

PART C: BIOLOGY THROUGH THE AGES

Your group should now refer to the set of *Biology Cards* at the end of the activity. Carefully read the information on each of the six cards. By discussing the events listed on each of the cards, decide as a group how the cards should be placed for them to be in *chronological* order (from the earliest events on Earth to the most recent events).

1. In what order did you place the *Biology Cards*?

	BIOLOGY CARD
1	
2	
3	
4	
5	
6	

2. As Earth evolved from one time period to the next, identify one change or event (piece of biological evidence) that enabled you to order the cards in the way you did.

TIME PERIOD TRANSITION	CRITICAL CHANGE OR EVENT
Hadean Eon → Archaean Eon	
Archaean Eon → Proterozoic Eon	
Proterozoic Eon → Paleozoic Era	
Paleozoic Era → Mesozoic Era	
Mesozoic Era → Cenozoic Era	

Activity 4

Match each of your *Biology Cards* with the corresponding time period from your *Timeline Cards* and *Geology Cards*. By placing the cards in order, you have created a biological timeline of events in Earth's past.

3. Get together with another group. Was your biological timeline the same as the group with which you compared your results? If not, how would you change your group's and/or the other group's timeline so they match? Describe the biological evidence that influenced any changes.

PART D: PUTTING IT ALL TOGETHER

There are strong cause-and-effect relationships between the geological and biological events that occurred throughout Earth's evolution. Using the timelines created in Parts B and C, work as a group to answer the following questions.

1. From the initial formation of Earth to the beginning of the Archaean eon, what changes or events were happening that prohibited ALL life from existing?

2. In order for prokaryotes to exist continuously on Earth, what geological changes or events had to have occurred first?

3. How did the development of early photosynthetic bacteria impact Earth's early atmosphere?

4. According to the geological evidence, when did this change to Earth's atmosphere caused by the photosynthetic bacteria first appear?

5. Did the biological activity from Question 3 and the geological evidence from Question 4 happen simultaneously? Or was there a delay between the biological event and the geological result? Explain why you think so.

The amount of protective ozone in Earth's atmosphere strongly influences the amount of damaging ultraviolet radiation that reaches Earth's surface. Based on this statement and the information on your *Geology* and *Biology Cards*, answer the following questions.

6. Before life could successfully exist on land, what significant change in Earth's atmosphere had to occur?

7. Prior to the change you identified in Question 6, what in Earth's environment did early bacterial life use to protect itself from the Sun's damaging UV radiation?

8. Prior to which time period would oxygen have been poisonous to early life-forms on Earth?

9. At the end of the Mesozoic era, a massive asteroid impact occurred on Earth. Using the information provided on your *Geology* and *Biology Cards*, answer the following questions.

    a. How might this asteroid impact have affected the world climate at this time?

    b. Did the impact result in a mass extinction of all life on Earth? Use evidence from your cards to explain your reasoning.

10. From the timeline your group completed in Part A, you learned that not all the time periods in Earth's history are the same length. Consider the events listed on the *Biology Cards*. Do shorter time periods or longer time periods demonstrate greater biological diversity? Support your answer with *specific* examples.

11. Now consider the events listed on the *Geology Cards*. Do shorter or longer time periods demonstrate more or less overall geologic change? Again, cite specific examples to support your response.

Activity 4

12. Periods of great geological activity correspond to periods of great biological diversity. Explain why you *agree* or *disagree* with this statement.

PART E: PUTTING IT IN PERSPECTIVE (OPTIONAL)

Now that we have explored the relationship between geological and biological events, let's put Earth's history into perspective. The process through which Earth formed and slowly evolved was extremely long and often tumultuous. During this part of the activity, you will explore a new way of thinking about the time it took for different events in Earth's history to occur.

1. The age of Earth is approximately 4,600 million years old (or 4.6 billion years old). As a group, create a relevant time scale (i.e. 1 hour is equivalent to some number of years) that will allow you to shrink the entire history of Earth into a single, 12-hour period (half a day). On this scale, the formation of planet Earth begins at midnight, and present-day Earth is at noon.

    Record your conversion factor here: _____ is equivalent to _____
                                                         hours   years

2. Using your time scale from Question 1, how many years will have gone by in 3 hours?

3. How many years will have passed in 6 hours?

4. Below you will find a series of boxes that contain clock readings. Each clock reading will represent a particular time in the 12-hour period and will correspond to a certain number of years before the present. Label each clock reading with the corresponding "million years ago" that is represented by the clock reading. Certain clock readings have been labeled for you.

    Note: The time periods represented by each clock reading will not directly correspond to the same time periods from your timeline in Part A. *One clock reading does not represent one card.*

5. Use your *Geology* and *Biology Cards* to identify only the events that had the greatest impact on the development of Earth. Since there are a greater number of hours on the clock than there are number of *Geology* or *Biology Cards*, you may decide to place events from the same card or time period next to more than one clock reading. Fill in each box with the geological and biological events that you identified from your cards. The cause-and-effect relationship between geological and biological events must be considered when approximating the order in which events occurred. In some cases we have filled in geological or biological evidence for you.

Evolving Earth

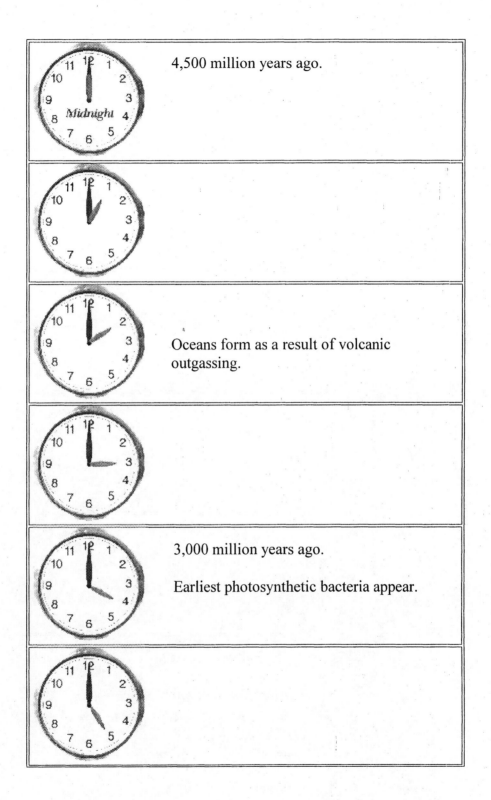

4,500 million years ago.

Oceans form as a result of volcanic outgassing.

3,000 million years ago.

Earliest photosynthetic bacteria appear.

Activity 4

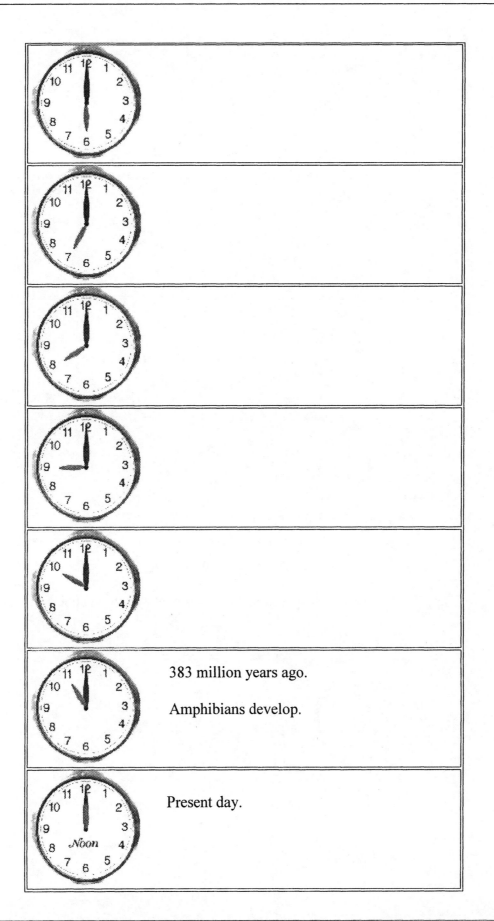

383 million years ago.

Amphibians develop.

Present day.

6. For how long (expressed in approximate hours and/or minutes) have humans been present on Earth?

7. Why do complex organisms appear so late in the half-day?

8. If life were discovered on a newly developing planet, what form of life (simple bacteria or complex) would you be most likely to find and why?

Timeline Cards

TIMELINE CARD

HADEAN EON

Time:

4,500 million years ago to 3,850 million years ago

TIMELINE CARD

ARCHAEAN EON

Time:

3,850 million years ago to 2,500 million years ago

TIMELINE CARD

PROTEROZOIC EON

Time:

2500 million years ago to 540 million years ago

TIMELINE CARD

PHANEROZOIC EON– PALEOZOIC ERA

Time:

540 million years ago to 250 million years ago

TIMELINE CARD

PHANEROZOIC EON– MESOZOIC ERA

Time:

250 million years ago to 65 million years ago

TIMELINE CARD

PHANEROZOIC EON– CENOZOIC ERA

Time:

65 million years ago to present

Geology and Biology Cards

GEOLOGY CARD

Earth's crust forms, oceans and atmosphere form as a result of volcanic outgassing

- Earth cools and solidifies
- Atmosphere rich in carbon dioxide and other gases toxic to today's complex life, but little oxygen
- Bombardment of Earth declines
- Continental plates form
- Plate tectonics begins
- Formation of the oldest rocks on Earth
- Early examples of sediments develop with banded iron formations
- Moon stabilizes tilt of Earth
- Volcanic activity high

BIOLOGY CARD

Life hits land

- Complex life develops
- Primitive plant life flourishes across the continents
- First vertebrates (fish) develop
- First oxygen-breathing animals evolve
- First amphibians, large insects, and reptiles develop

BIOLOGY CARD

Rise of multicellular organisms

- Eukaryotes develop
- Multicellular life develops
- Plants like seaweed and kelp develop and anchor to the ocean floor
- Primitive anaerobic microorganisms undergo significant population reduction due to oxygen poisoning
- Life still restricted primarily to oceans

GEOLOGY CARD

Formation of Earth

- No stable atmosphere
- Earth continuously bombarded by planetesimals and meteors
- Formation of central iron core
- Moon formed by giant impact with Earth
- Jupiter stabilizes Earth's orbit

Life in the Universe – Activities Manual, 2nd Edition   Prather, Offerdahl, and Slater

Geology and Biology Cards

GEOLOGY CARD

Continents divide

- Pangea breaks up
- Global climate warms
- Poles are free of ice
- Sea levels rise, submerging 30% of present-day continental mass
- Coal and petroleum deposits formed by decaying plants and animals
- Period ends with massive asteroid impact, inserting Sun-blocking debris into the atmosphere that forces climate into cold, dark ice age

BIOLOGY CARD

Bacterial life forms appear

- Prokaryotes appear
- Earliest bacteria obtain energy through ingestion of organic molecules and/or geochemical disequilibria
- Predominantly anaerobic microorganisms (those that flourish only in low-oxygen environments) dominate Earth
- First photosynthetic bacteria (cyanobacteria) appear and their biological activity begins to supply oxygen to atmosphere
- All life restricted to the oceans.

BIOLOGY CARD

Age of mammals

- Dinosaurs are extinct
- Mammalian life diversifies
- Rise of mammals and grazing animals
- Rise of first primates
- Rise of humans

GEOLOGY CARD

Modern mountain chains form

- Most recent glaciation, where ice covers much of northern United States
- Grand Canyon carved
- Plate tectonics forms Himalayas and Cascades
- Continents move into present-day position
- Mild climate

Geology and Biology Cards

GEOLOGY CARD

Plate tectonics slows

- Volcanic activity slows
- Oxygen builds up in atmosphere and in oceans to near current-day levels
- Free oxygen forms ozone layer
- Continent building and continental drift due to plate tectonics
- End of the banded iron formations due to increased atmospheric oxygen

BIOLOGY CARD

No life

- The environment of Earth is too hostile to support life

BIOLOGY CARD

Age of dinosaurs

- Early dinosaurs evolve
- First birds appear
- Dinosaurs dominate land
- Many land plants, ferns, primitive trees
- Large fish and sharks in oceans
- Flowering plants evolve
- Animals and plants spread across all continents
- Another mass extinction occurs at the end of this time period

GEOLOGY CARD

Continents merge

- Continental plates and land masses combine to form supercontinent Pangea
- Climate mild, variable, and moist
- Extensive areas of glaciation and desert exist
- Water cycle drives erosion
- Rivers and lakes form
- Mountains build on continents
- Volcanic activity slows further

Life in the Universe – Activities Manual, 2nd Edition   Prather, Offerdahl, and Slater

5
THE NATURE OF LIFE

GOALS
- Develop a definition of life
- Recognize the difficulties in establishing an absolute definition of life
- Discuss the impacts that our knowledge of life on Earth has on the search for life on other planets and moons

As we begin the search for life beyond Earth, it is important to understand life on our own planet. What constitutes life? Is it easily identified? What are the characteristics of life on Earth? In this activity, you will explore the definition of life and how this definition impacts how we search for life in the universe.

PART A: IS IT LIVING? - IDENTIFYING THE DEFINING CHARACTERISTICS OF LIFE

Obtain two plant specimens from your instructor. In this part of the activity, you may use all of your senses - touch, taste, smell, sight, and sound - to make observations about each of the plant specimens. Use only your senses to answer the following questions.

1. As a group, determine which plant specimen is living. List more than one observable trait of the specimen that supports the conclusion that it is *alive*. Remember that you may only cite evidence gathered by the use of your five senses.

2. What evidence leads you to the conclusion that one of the specimens is *not* living? List more than one observable trait of this specimen to support your conclusion.

3. List any characteristics the nonliving specimen has in common with the living specimen. How did this impact your decision about which is alive and which is not?

Activity 5

4. If you were a scientist and had access to the necessary equipment and supplies, what tests would you run or what additional evidence (beyond that gained by your five senses) would you seek to support your choice from Question 1? Why?

The definition of life used by scientists is continually changing as we gain more evidence and make new discoveries. Throughout this activity, we will explore different situations that will require you to develop your own functioning definition of life, which may or may not change as you make additional observations.

5. Based on your experiences so far in this activity, list what you feel are the basic defining characteristics of life.

6. Compare your list from Question 5 with another group.

   a. How was your list different from or similar to the group with which you compared your list?

   b. Do you agree with their list of the basic characteristics for life? If not, why not?

   c. Are there any changes you would make to your list? If yes, what are they and why would you consider making each of these changes? If not, why not?

PART B: LOOK BUT DON'T TOUCH!

Obtain three Petri dishes or cups, labeled A, B, and C, from your instructor. Each Petri dish contains a potentially living specimen. You may examine the contents of each Petri dish but cannot touch its contents. The contents of the Petri dish must remain inside the dish at all times.

1. Predict which Petri dish(es) your group thinks contains a living specimen. Cite your evidence, if any, in each case to support your prediction.

    Petri Dish A:

    Petri Dish B:

    Petri Dish C:

2. Gently add lukewarm water to each of the three Petri dishes. Record any changes and/or observations below.

    Petri Dish A:

    Petri Dish B:

    Petri Dish C:

3. Based on your observations from Questions 1 and 2, which Petri dish (or dishes) contains a living specimen? Explain your reasoning.

    Petri Dish A:

    Petri Dish B:

    Petri Dish C:

Activity 5

4. Your instructor will now reveal to the class which Petri dish (or dishes) contains a living specimen.

    a. Were you correct about which Petri dish (or dishes) truly contains a living specimen? If not, what observations or evidence led you to the wrong conclusion?

    b. What other observations could you have made or tests could you have performed to determine which specimens are alive?

5. Did any of your observations or conclusions from Questions 1 through 4 influence your list of basic defining characteristics of life from Part A? If yes, which characteristics were they? How would you change your list of characteristics?

PART C: THE TOUGH CASES

Although there is no single definition, a simplified list of six requirements that are commonly used to define life includes the abilities of an organism to (1) reproduce faithfully, (2) utilize an energy source, (3) grow and develop, (4) respond and adapt to the surrounding environment, (5) exhibit order in their internal structure, and finally (6) evolve. However, as you may have discovered in Parts A and B, it is much more difficult to develop an absolute definition of life. In this section, we will continue to explore the definition of life and how it impacts the search for life on other planets.

1. Use your list of defining characteristics for life as well as the six requirements above to reason about the following situations.

    a. A mule is the offspring of a donkey and a horse but is unable to reproduce. Is it still considered to be alive? Why or why not?

b. A rock crystal has the ability to spontaneously grow and get larger like many organisms by repeating its crystalline structure. Would your group consider rock crystals to be living or nonliving? Explain your reasoning.

c. Fire has the ability to utilize an energy source and to grow. Would your group consider fire to be living or nonliving? Explain your reasoning.

d. Would a robot that could operate a fabrication plant to make copies of itself be alive? Why or why not?

e. Viruses have the ability to infect complex organisms in the same way as bacteria do and oftentimes produce similar results. However, viruses lack much of the cellular "machinery" for reproduction and certain metabolic processes. Therefore, viruses cannot reproduce without the assistance of a host cell. Parasites are another type of organism that also live off a host organism, either by living within or on the host organism. Based on this, your definitions of life, and any other outside knowledge, do you feel viruses are living? Explain your reasoning thoroughly.

2. If you were to randomly land on Earth, you might end up at a location in which you would not be able to detect life (without using sophisticated instruments).

   a. Give one or two examples of such a location and explain why you would not be able to detect life there.

Activity 5

   b. If an alien landed on Earth in any of the locations listed in Question 2a, would the determination they make from that location correctly depict whether or not life exists on all of Earth? Why or why not?

3. Of all the data we have collected about the surface of Mars (from the *Voyager*, *Pathfinder*, *Spirit*, and *Discovery* missions), can we accurately determine whether or not life exists on all of Mars? Why or why not?

4. As you may have discovered in doing this activity, the definition of life on Earth is not absolute. How does this complicate the search for life on other planets?

5. In our search for life on other planets or moons beyond Earth, should we be looking for life that is similar to life-forms on Earth? Why or why not?

6
DESIGNER GENES FOR A DESIGNER WORLD

GOALS
- Explore the interactions between hypothetical extremophiles and extreme environments
- Examine the role genetic mutations play in adaptation
- Practice writing and interpreting fictitious genetic codes
- Consider extreme environments throughout the solar system and the possibility of life within these environments

PART A: CREATE YOUR OWN EXTREMOPHILE

With growing speculation about life on other planets, astrobiologists have turned their attention to extreme environments on Earth and the organisms that live there. The term *extreme environment* most often describes any environment that is extreme with respect to human tolerances. The single-celled microorganisms thriving in such environments are called extremophiles. In this portion of the activity, you will create both extremophiles and extreme environments using two lists of characteristics. Your group will then pair each imaginary extremophile randomly with a different environment.

1. As a group, decide which characteristics you want your extremophile to possess. To do this, make a single choice from each of the five categories of *Extremophile Characteristics* listed below. Once you have chosen five characteristics in total, fill out one of the *Extremophile Cards* found at the end of the activity. Repeat this process to construct two more extremophiles and cards. Make sure you create significantly different extremophiles each time.

2. Follow the steps discussed in Question 1 by using the five categories of *Environmental Characteristics* to create an interesting environment. (Note: The environments you create in no way need to match the extremophiles you have already created.) Record your choices on your *Environment Cards* located at the end of this activity. Complete three distinct *Environment Cards*.

Extremophile Characteristics

A. RESPIRATION
- Uses oxygen
- Uses sulfates/sulfur
- Uses nitrates
- Uses carbonates
- Uses iron

B. TEMPERATURE PREFERENCE
- Extremely cold (-3°C to 20°C, 26°F to 68°F)
- Moderate (20°C to 35°C, 65°F to 95°F)
- Extremely hot (42°C to 113°C, 107°F to 236°F)

C. ENERGY SOURCE
- Photons (light)
- Organic carbon (simple sugars, carbohydrates, etc.)

Activity 6

- Geochemical energy (involving inorganics)

D. **pH PREFERENCE**
- Low (0 to 5)
- Medium (5 to 9)
- High (9 to 14)

E. **OTHER**
- Prefers environments with extreme pressures (> 100 atm)
- Prefers environments with low pressure (< 0.1 atm)
- Prefers environments with high concentrations of salt (> 5%)
- Prefers environments with low concentrations of salt (< 5%)
- Thrives in environments with extremely small amounts of water/moisture

Environmental Characteristics

A. **RESPIRATION**
- Abundant oxygen
- Abundant nitrates
- Abundant sulfates/sulfur
- Abundant carbonates
- Abundant iron

B. **TEMPERATURE**
- Low temperatures (-3°C to 20°C, 26°F to 68°F)
- Moderate temperatures (20°C to 35°C, 65°F to 95°F)
- High temperatures (42°C to 113°C, 107°F to 236°F)

C. **ENERGY SOURCE**
- Abundant light
- Organic carbon highly accessible
- Source of geochemical energy accessible

D. **pH LEVEL**
- Low pH (0 to 5)
- Moderate pH (5 to 9)
- High pH (9 to 14)

E. **OTHER**
- High barometric pressures (>100 atm)
- High concentration of salt (> 5%)
- Little to no liquid water/moisture present
- Low barometric pressures (< 0.1 atm)
- Low concentration of salt (< 5%)

3. Once your group has completed three *Extremophile Cards* and three *Environment Cards*, remove the card pages from your activity book and separate the cards. Keep the *Extremophile Cards* separate from the *Environment Cards*. Your instructor will collect each group's cards, placing all of the *Extremophile Cards* from the class in one pile or box and all of the *Environment Cards* from the class in another pile or box.

4. Draw three new *Extremophile Cards* and three new *Environment Cards*. Read the information on your new *Environment Cards* carefully. Would humans be able to live in the conditions provided? If not, what about these environments is extreme to humans?

5. Randomly choose one of your *Extremophile Cards* and pair it with one of your *Environment Cards*. Repeat this process again until you have created three pairs, each pair consisting of one *Extremophile Card* and one *Environment Card*.

Your group should now have three distinct extremophile/environment pairs to study. In this portion of the activity, you will be the judge of the success of each extremophile in the environment with which it was paired. To do this, you will compare each extremophile characteristic to the corresponding environmental characteristic. To score your extremophiles, use the following set of guidelines.

   a. Award one point if the extremophile characteristic helps the extremophile survive in the environment with which it was paired. For example, if the extremophile characteristic chosen from the "Respiration" category is "Uses oxygen" and the corresponding environmental characteristic from the "Respiration" category is "Abundant oxygen," then that extremophile earns one point.

   b. Award zero points if there is not enough information to determine if the extremophile characteristic and environmental characteristic have a positive, negative, or null effect on the success of the extremophile. For example, if the extremophile characteristic chosen from the "Other" category is "Prefers environments with high concentrations of salt" and the corresponding environmental characteristic chosen from that category is "Low barometric pressures," then that extremophile earns zero points for that category.

   c. Subtract one point if the environmental characteristic restricts the extremophile's ability to survive. For example, if the extremophile characteristic chosen from the "Respiration" category is "Uses oxygen" and the corresponding environmental characteristic from the "Respiration" category is "Abundant carbonates," then that extremophile loses one point.

6. Score each of your three pairings of extremophile and environmental characteristics. Record all scores, including the total score for each extremophile, in the spaces below.

Pairing #1

Category	Score
Respiration	
Temperature Preference	
Energy Source	
pH preference	
Other	
TOTAL SCORE:	

Pairing #2

Category	Score
Respiration	
Temperature Preference	
Energy Source	
pH preference	
Other	
TOTAL SCORE:	

Pairing #3

Category	Score
Respiration	
Temperature Preference	
Energy Source	
pH preference	
Other	
TOTAL SCORE:	

Activity 6

7. Would you rate the extremophile with the greatest score as being very successful, moderately successful, or not successful in its environment? While examining each type of characteristic, describe the success or failure of this extremophile in terms of the interactions between the extremophile and its environment.

8. Give one example of an organism found on Earth that has a unique characteristic that makes it specifically suited to live in a particular environment.

9. Sometimes the characteristic that makes an organism successful in its natural environment becomes useless or detrimental when the organism is placed in another environment. Describe a real organism and environment found on Earth that would be an example of this situation.

10. What *one* change would you make to the extremophile from Question 7 to make it more successful in the environment with which it was paired? Explain why you chose to make this change. If you feel no changes are needed for this extremophile, choose one of your less successful extremophiles to answer this question.

PART B: PLAYING WITH GENES

Our physical characteristics are, to a large degree, due to the genes we inherit from our parents. Our genes are simply strings of four different molecules called *nucleotides*. We often represent these nucleotides as single letters A, C, G, and T. The order of nucleotides in a gene determines the structure of a protein.

Proteins are the building blocks and workhorses of cells. As such, some proteins act as the "bricks" of our cells while others facilitate the chemical reactions needed for cells to survive in their environment. Without the correct set of proteins to speed up cellular reactions in particular environments, cells would die. For example, cells exposed to high temperatures must have proteins that hold cell machinery together under extreme temperatures. Or cells living in extremely cold temperatures increase the amount of salt inside the cell to act as antifreeze, preventing the cell from freezing solid.

Mutations occur when a permanent change in the sequence of nucleotides in a gene is made. A change in nucleotide sequence alters the gene which can then result in a different protein structure and/or function, thereby enabling or preventing a cell to live in a particular environment.

Designer Genes

In this part of the activity, we will experiment with a made-up set of nucleotides: W, X, Y, and Z. For our purposes, we will imagine that a string of four of these nucleotides represents one gene that determines the structure of a single protein. In our model, a protein then impacts the physical characteristics of an organism.*

1. As is often the case in nature, extremophiles may exist in a combination of extreme environmental conditions. For example, extremophiles that live in the sulfuric hot springs of Yellowstone National Park are both thermophiles and acidophiles and are therefore called thermoacidophiles. Below we describe an extremophile that is able to thrive in a variety of extreme environmental conditions. Using the table below, write out the nucleotide sequence that would make proteins resulting in the physical characteristics of this extremophile.

    Extremophile #1: Uses oxygen, lives in moderate temperatures, uses photons for energy, prefers moderate pH, and prefers environments with extreme pressure.

Nucleotide Sequence:

CHARACTERISTIC	NUCLEOTIDE SEQUENCE
Uses oxygen	WWXX
Uses nitrates	XXZZ
Uses sulfates/sulfur	YYXX
Uses iron	ZZYY
Uses carbonates	WZXY
Lives in very cold environments	XZYX
Lives in moderate environments	WYXY
Lives in very hot environments	YXWX
Uses photons for energy	YZWW
Uses organic carbon for energy	WXYW
Uses geochemical energy (involving inorganics)	ZZYZ
Prefers low pH	YYXY
Prefers moderate pH	XZXX
Prefers high pH	XWZZ
Prefers environments with extreme pressure	XYZZ
Prefers environments with low pressure	ZWYY
Prefers environments with high salt concentration	WWWW
Prefers environments with low salt concentration	XXXX
Thrives in environments with extremely small amounts of water/moisture	ZZZZ

Activity 6

It is important to note that in reality several thousand to several million nucleotides represent one gene and that physical characteristics are often determined by one or several genes just as a single gene can influence one or several proteins which can impact one or more characteristics. In this activity, we will only examine mutations that result in the change of a single characteristic of the organism. We will assume that each characteristic of an organism can be represented by a single protein (encoded by a gene or sequence of four nucleotides).

2. Imagine that Extremophile #1 from Question 1 lives around geothermal vents on the ocean floor (Environment #1) which has the following environmental conditions:

    Environment #1: Oxygen present, high temperatures, organic carbon present, low pH, and extreme pressure.

    Based on this information, score the success of Extremophile #1 in Environment #1 using the scoring rules from Part A.

3. Do you think that Extremophile #1 would thrive in Environment #1? Explain your reasoning in detail.

Now suppose that over a long period of time the nucleotide sequence of Extremophile #1 developed mutations that eventually resulted in the following characteristics in several of its offspring (Extremophile #2):

   Extremophile #2: Uses oxygen, lives in very hot environments, uses organic carbon for energy, prefers low pH, and prefers environments with extreme pressure.

4. Write out the nucleotide sequence for Extremophile #2.

5. Score the success of Extremophile #2 in Environment #1 (from Question 2).

6. Would Extremophile #2 be more successful than Extremophile #1 in Environment #1? Explain your reasoning.

7. Based on your response to the previous question, do you feel that the mutations in the nucleotide sequence for Extremophile #2 were advantageous to the organisms or disadvantageous? Why?

Now suppose that over a long period of time Environment #1 changes to have the following conditions:

Environment #2: Oxygen present, very cold temperatures, no organic carbon present/intermittent photon energy present, moderate pH, very high concentration of salt.

8. Would Extremophile #1 or Extremophile #2 be more successful in Environment #2? Explain your reasoning.

9. In general, under what circumstances would mutations be considered advantageous and under what circumstances would mutations be considered disadvantageous?

10. In Questions 4 through 8, you identified several mutations within a nucleotide sequence that, when expressed by an extremophile, impacted its ability to survive an environmental change. What had to be passed on in order for these mutations to be expressed in the offspring of Extremophile #1?

11. As an environment changes over an extended period of time, we note that a *population* of organisms may exhibit different physical characteristics that make the species better adapted to those environmental changes. What aspect of the organisms has changed resulting in a better chance of survival?

12. Suppose a mutation occurs within the genetic code of a single extremophile and results in a characteristic change that is advantageous to the organism. How long would it take for the mutation to be expressed in the majority of a *population* of extremophiles? Choose either a, b, or c and explain in detail why you chose the one you did.

    a. Within the lifetime of the original extremophile.

Activity 6

    b. After a single lifetime of the offspring of the original extremophile.

    c. After several generations of extremophiles.

13. Microorganisms reproduce *much* more quickly than multicellular organisms such as plants and animals. Would it take more time, less time, or an equal amount of time for a change to be expressed in an entire population of microorganisms (such as extremophiles) as compared to an entire population of multicellular organisms?

PART C: EXTREME ENVIRONMENTS IN OUR SOLAR SYSTEM

Now that we have experimented with a variety of hypothetical extreme environments on Earth, lets examine some of the extreme environments found in our solar system. Below is a list of various environmental characteristics for Titan, one of Saturn's moons. Next to it is a hypothetical extremophile. Use this information to answer the following questions.

TITAN'S ENVIRONMENT	HYPOTHETICAL EXTREMOPHILE
Characteristics	**Characteristics**
Atmosphere containing nitrogen, methane, ethane, argon, and hydrogen	Thrives in environments with low moisture/water
	Prefers low pH
Very cold temperatures	Uses oxygen
Possible lakes/oceans of ethane/methane	Uses organic carbon for energy
Abundant light	Prefers very cold environment
May have continents	

1. Using the same scoring criteria from Part A, what score would this extremophile receive if it were placed on Titan? How would you rate the prospects for success of this extremophile in Titan's environment?

2. What characteristics of the extremophile would you change in order to make it more successful in this environment?

3. What aspect of this population of extremophiles would need to change over time in order for later generations to be more successful?

4. Now, utilizing the set of *Planet/Moon Cards* from the end of this activity, design an extremophile that could live in each of the planet/moon environments described on your cards. Fill out an *Extremophile Card* for each new extremophile and pair it with the corresponding planet or moon.

5. Using the table of nucleotide sequences shown earlier, write out the genetic makeup for each of your new extremophiles.

    Mars Extremophile:

    Europa Extremophile:

    Io Extremophile:

6. The total amount of genes in a cell is called the *genome*. Typical mammalian genomes have up to 750 times as much genetic information as bacterial genomes. Which has more nucleotide sequences, the mammalian genome or the bacterial genome?

7. Which expresses a wider variety of characteristics, mammals or bacteria?

8. If you were to change two genes in a bacteria and two genes in a mammal, which organism would demonstrate a greater overall difference in the total characteristics expressed? Explain why you think so.

9. Based on your answer to Question 8, would you expect a bacteria or a mammal to adapt more *easily* to the same environmental changes? Describe why you think this.

10. Based on your answer to Question 9, which would you expect to adapt more *quickly* to the same environmental changes, a population of bacteria or a population of mammals? (Note: Keep in mind your response from Part B, Question 7.)

Activity 6

11. If we were looking for life on another planet or moon in the universe, do you think it would be more likely to discover microbial life or more complex life? Explain your response in detail.

12. Which of the planets or moons on your cards do you think is most likely to already have living organisms on it? Explain your reasoning.

13. On which of the planets or moons do you think Earth-like life-forms could survive? Explain why you think this.

Extremophile and Environment Cards

EXTREMOPHILE CARD

	Characteristic
Category 1: Respiration	
Category 2: Temperature Preference	
Category 3: Energy Source	
Category 4: pH Preference	
Category 5: Other	

EXTREMOPHILE CARD

	Characteristic
Category 1: Respiration	
Category 2: Temperature Preference	
Category 3: Energy Source	
Category 4: pH Preference	
Category 5: Other	

EXTREMOPHILE CARD

	Characteristic
Category 1: Respiration	
Category 2: Temperature Preference	
Category 3: Energy Source	
Category 4: pH Preference	
Category 5: Other	

ENVIRONMENT CARD

	Characteristic
Category 1: Respiration	
Category 2: Temperature	
Category 3: Energy Source	
Category 4: pH Level	
Category 5: Other	

ENVIRONMENT CARD

	Characteristic
Category 1: Respiration	
Category 2: Temperature	
Category 3: Energy Source	
Category 4: pH Level	
Category 5: Other	

ENVIRONMENT CARD

	Characteristic
Category 1: Respiration	
Category 2: Temperature	
Category 3: Energy Source	
Category 4: pH Level	
Category 5: Other	

Life in the Universe – Activities Manual, 2nd Edition — Prather, Offerdahl, and Slater

Extremophile and Planet/Moon Cards

EXTREMOPHILE CARD	
	Characteristic
Category 1: Respiration	
Category 2: Temperature Preference	
Category 3: Energy Source	
Category 4: pH Preference	
Category 5: Other	

EXTREMOPHILE CARD	
	Characteristic
Category 1: Respiration	
Category 2: Temperature Preference	
Category 3: Energy Source	
Category 4: pH Preference	
Category 5: Other	

EXTREMOPHILE CARD	
	Characteristic
Category 1: Respiration	
Category 2: Temperature Preference	
Category 3: Energy Source	
Category 4: pH Preference	
Category 5: Other	

Mars

Environmental Characteristics
Carbon dioxide atmosphere
Abundant sunlight
Very cold temperatures
Many carbonates in soil
Organics abundant in soil

Europa

Environmental Characteristics
Thin oxygen atmosphere
Very cold temperatures
Lots of ice present
Water may be present
Limited sunlight

Io

Environmental Characteristics
Thin sulfur dioxide atmosphere
Very cold temperatures
No water
Abundant light
Volcanically active/possible sulfur lakes

7
THE EXTREME ENVIRONMENTS OF EARTH AND THE CREATURES THAT LIVE THERE

GOALS
- Explore the environmental conditions that limit life on Earth
- Identify the characteristics of different extremophiles
- Explore optimum environmental conditions in which extremophiles thrive
- Consider cause-and-effect relationship between environment and organism success within that environment
- Learn about extreme environments within our solar system and reason about the possibility of life within those environments

PART A: INTRODUCTION TO EXTREMOPHILES

1. As a group, *describe* environments on Earth that you believe would not allow any form of life to exist. List specific examples, and explain *why* these environments cannot support life.

2. After surveying a large number of college and high school aged students, it was discovered that the three most common environments listed as being unable to support life were (1) environments in which there is no oxygen, (2) environments with extreme temperature (too hot/too cold), and (3) environments without access to water.

   a. Did your group include any of these environments? Which ones?

   b. If not, do you feel that you should have? Explain why or why not.

3. Scientists are currently searching for life on other planets and moons both within and outside our solar system. What types of life-forms do you think they are most likely to find: (1) animals, (2) plants, or (3) microorganisms? Explain your reasoning.

Activity 7

4. Refer to the pages titled *Extremophile Data Sheet* found at the end of this activity. Carefully read through the information provided. Use this information to complete the *Extremophile Data Table* also found at the end of this activity.

5. Now that you have read the information sheets and completed the chart, refer back to the environments listed in Question 2. Do you think that any of these environments *actually* present conditions too extreme for life to exist? Explain why you think so.

6. During class, you overhear two students discussing extremophiles. Carefully read the discussion below and answer the following question.

    Student #1: Xerophiles are extremophiles that prefer environments that are very dry. They can live in places that never have any water.

    Student #2: I disagree. Although xerophiles do thrive in extremely dry environments, they need at least a little water part of the time in order to survive. I think all life as we know it needs some water to live.

    Do you agree or disagree with either or both of the students? Explain your reasoning for each.

7. What is the highest temperature in which extremophiles are known to live? What do we call the extremophiles that prefer living in these temperatures?

8. What is the lowest temperature in which extremophiles are known to live? What do we call the extremophiles that prefer living in these temperatures?

9. pH describes how acidic or how basic the environment is. What are the names of the two types of extremophiles that can live in each of the conditions of extreme pH? What are the pH ranges in which they live?

10. What is the one environmental condition that *all life* needs in order to survive on Earth?

PART B: WHO LIVES WHERE?

In this activity you will investigate three hypothetical environments and three bacterial life-forms. For each environment and bacteria, we have provided a table (on the following page) with a partial list of characteristics that describe (1) how the different environments support life and (2) the different needs of each bacteria that allow it to thrive in a particular environment. You will examine the characteristics that are provided for each environment and bacteria and then, based on this information, complete each table by deciding which type of bacteria can live in each of the environments.

In the first column of the table below we have listed the essential characteristics that you will need to consider when trying to match a bacteria and its environment. In the column next to each characteristic are the possible ranges of values for each characteristic.

CHARACTERISTICS	RANGE OF VALUES
Temperature	$0°C - 100°C$
Salinity	Low, Medium or High <5% to 30%
pH level	1 – 14 (pure water has a pH of 7)
Energy Sources/Requirements	Sunlight (photons) or Chemical Potential (energy)
Carbon Sources/Requirements	Organic (sugars/proteins/fats) or Inorganic (CO_2 or CO_3^{2-})
Oxygen Provided by the Environment or Used by Bacteria	Yes or No

Activity 7

1. First, decide which bacteria type (A, B, or C) matches each environment (X, Y, or Z). Only one type of bacteria matches each environment. Once you have matched each type of bacteria with a single environment, completely fill in all the blanks in each of the tables with the corresponding information provided.

Environment X

Temperature	2°C
Salinity	12%
pH Level	
Energy Source	Light-Photons
Carbon Source	Inorganic
Oxygen Provided	

Environment Y

Temperature	90°C
Salinity	3%
pH Level	
Energy Source	
Carbon Source	Inorganic
Oxygen Provided	No

Environment Z

Temperature	25°C
Salinity	22%
pH Level	
Energy Source	
Carbon Source	Organic
Oxygen Provided	No

Bacteria Type A

Preferred Temperature	20°C – 30°C
Preferred Salinity	
Preferred pH Level	7
Energy Source Used	Chemical Potential
Carbon Source Used	
Oxygen Needed	

Bacteria Type B

Preferred Temperature	
Preferred Salinity	
Preferred pH Level	7
Energy Source Used	
Carbon Source Used	CO_2
Oxygen Needed	Yes

Bacteria Type C

Preferred Temperature	
Preferred Salinity	
Preferred pH Level	3
Energy Source Used	Chemical Potential
Carbon Source Used	CO_2
Oxygen Needed	

Extreme Environments

2. State which bacteria type (A, B, or C) you decided could live in which environment (X, Y, or Z).

3. Now that you have completed each of your environment cards, state each of the conditions that make it an extreme environment. It is not necessary to consider the energy source or the carbon source in this description. (Note: You may want to revisit the definition of extreme environments in the first paragraph of your *Extremophile Data Sheet*.)

   Environment X: Environment Y: Environment Z:

4. For bacteria type A, which of the conditions of the other two environments made them too extreme for bacteria type A to survive? (In what ways does bacteria type A not match these environments?)

   Environment __: Environment __:

5. For bacteria type B, which of the conditions of the other two environments made them too extreme for bacteria type B to survive? (In what ways does bacteria type B not match these environments?)

   Environment __: Environment __:

6. For bacteria type C, which of the conditions of the other two environments made them too extreme for bacteria type C to survive? (In what ways does bacteria type C not match these environments?)

   Environment __: Environment __:

Activity 7

7. If the number of photons that arrive at environment Y were to decrease to nearly zero, would the bacteria that you chose still be able to thrive in this environment? Explain why or why not.

8. Would your answer to Question 7 change if we were instead considering environment X or Z and the corresponding bacteria? Explain your reasoning.

9. If the bacteria you chose to pair with environment Z was placed in an environment without oxygen, would the bacteria that you chose still be able to thrive? Explain why or why not.

10. Would your answer to Question 9 change if we were instead considering the bacteria you chose to pair with environment X or Y? Explain your reasoning.

11. Are the bacteria you considered in Questions 9 and 10 aerobic bacteria or anaerobic?

12. Which type of bacteria (A, B, or C) uses a carbon source that is organic and which type uses a source of inorganic carbon?

13. Which of these hypothetical bacteria (A, B, or C) is a:

    Thermophile: ___

    Psychrophile: ___

    Halophile: ___

Extreme Environments

PART C: THE SEARCH FOR LIFE IN THE UNIVERSE

1. Refer to the *Extreme Environments Beyond Earth* data sheet at the end of the activity that lists extreme environments in our solar system. In the space provided, state whether or not each of these planets or moons would provide an environment in which an extremophile might exist. In each case, describe the extreme environment(s) and corresponding extremophile(s) that may be able to survive under these conditions.

   a. Mars

   b. Venus

   c. Europa

   d. Titan

   e. Io

2. Consider your responses to Question 1. If you were looking for life in our solar system, which planet or moon would you consider investigating first? Explain your reasoning in detail.

3. What changes in the atmosphere or surface characteristics would most strongly increase the chance for life to exist on the planet or moon you selected in Question 2?

Activity 7

4. During class, your professor tells you that many Jupiter-sized planets have been discovered orbiting other stars in our galaxy. She describes that some of these gas giant planets are close enough to their companion stars that the planet's average temperature could be high enough for the presence of liquid water. You subsequently overhear two students discussing the possibility of finding life in the galaxy. Carefully read the discussion below and answer the following question.

   Student #1: I don't think there's any chance life can be found on any of these newly discovered gas giants because there is no water or surface on which life could form. Discovering Jupiter-sized planets won't ever lead to us finding life outside our solar system.

   Student #2: I disagree. It's not about the possibility of finding life on the Jupiter-sized planets. Big planets like Jupiter can also have lots of moons orbiting them. That is where we might find life.

   Do you agree or disagree with either or both of the students? Explain your reasoning for each.

Extremophile Data Sheet

Can you imagine living in an environment as salty as the Dead Sea? Or how about a place where the temperature is continually below the freezing point of water? Could you survive in an environment without oxygen? The term *extreme environment* is used to describe any environment that is extreme with respect to human tolerances. Human existence would be difficult at best in any of these conditions. However, scientists have found that life, in some form, exists in virtually every environment on Earth. Most of this life is in the form of microorganisms. Microorganisms that thrive in extreme environments are called *extremophiles*. Extremophiles are further classified according to the environments in which they thrive. Examples of specific extremophile classifications are thermophiles, psychrophiles, acidophiles, alkalophiles, barophiles, xerophiles, halophiles, and anaerobes. Listed below is information about each of these extremophiles, including optimum conditions for existence as well as examples of specific microorganisms.

Thermophiles (*hot-loving*):

Preferred Environmental Conditions

- These microorganisms not only survive in areas of high temperature, they thrive there
- Temperature range: $42°C$ to $121°C$ ($107°F$ to $236°F$)
- Thrive in temperatures well over the boiling point of water

Preferred Environments

- Bottom of the ocean near geothermal vents
- Hot springs in Yellowstone National Park

Examples of Thermophilic Microorganisms

- *Thermus aquaticus* – organism famously used in polymerase chain reaction (PCR) that enables geneticists to amplify DNA
- *Pyrolobus fumerii* – microorganism that has an optimum growth temperature of $106°C$, although it has been found in nature at temperatures of up to $113°C$
- Archaea "Strain 121" – so named because it was found in water at $121°C$

Psychrophiles (*cold-loving*):

Preferred Environmental Conditions

- Microorganisms that live in temperatures that linger around the freezing point of water and sometimes lower
- Temperature range: $-20°C$ to $20°C$ ($26°F$ to $68°F$)

Preferred Environments

- Antarctic snowfields
- Surface of glaciers
- Deep in ocean away from hydrothermal vents (constant temperature of $1°C$ to $3°C$)

Extremophile Data Sheet Continued

Examples of Psychrophilic Microorganisms

- *Heteromita globosa* – found in the snowfields in Antarctica
- *Chlamydomonas nivalis* – snow algae known to turn the surface of glaciers pinkish-red

Acidophiles (*acid-loving*):

Preferred Environmental Conditions

- Microorganisms that thrive in extremely acidic conditions; these critters actually prefer to live in liquids similar to sulfuric acid
- pH range: less than pH 5

Preferred Environments

- Sulfur hot pots of Yellowstone National Park
- Volcanic soils
- Gastric fluids of the stomach

Examples of Acidophilic Microorganisms

- *Cyanidium caldariam* – organism that can grow (albeit slowly) in a pH as low as zero
- The *Sulfolobus* species – lives in a preferred pH of less than 2, found in Yellowstone hot pots

Alkalophiles (*base-loving*):

Preferred Environmental Conditions

- Microorganisms that prefer to live in very alkaline environments
- pH range: greater than pH 9

Preferred Environments

- Soda lakes in Africa
- Soils with high-carbonate soils

Examples of Alkalophilic Microorganisms

- *Spirulina* – a cyanobacteria found living in a soda lake with pH ~10

Extremophile Data Sheet

Barophiles (*pressure-loving*):

Preferred Environmental Conditions

- These microorganisms thrive under high pressure
- Pressure range: pressures hundreds of times greater than 1 atm (Note: We exist surrounded by a pressure of about 1 atm at sea level.)

Preferred Environments

- Deep ocean environments
- Equipment sterilized using high pressure

Examples of Barophilic Microorganisms

- Entire eukaryotic communities on the continental shelf
- Sea cucumbers
- Microorganisms around deep sea black smokers

Xerophiles (*dry-loving*):

Preferred Environmental Conditions

- Microorganisms that grow in *extremely* dry environments with only intermittent moisture

Preferred Environments

- Deserts
- Antarctica dry valleys
- Cereals, dried fruit, candy
- On stones

Examples of Xerophilic Microorganisms

- *Xeromyces bisporus* – xerophilic fungi living on dried foods
- *Metallogenium* – a genus of desert bacteria that produce magnetite

Halophiles (*salt-loving*):

Preferred Environmental Conditions

- Microorganisms that thrive in environments with high concentrations of salt
- Salt concentration: >5% for optimum growth (sea water is about 3%)

Extremophile Data Sheet Continued

Preferred Environments

- Soda and salt lakes or bodies of water (e.g., Great Salt Lake)
- Salted fish and meats (beef jerky)

Examples of Halophilic Microorganisms

- *Halobacterium* – can live in salt concentrations of up to 30%
- *Staphylococcus* – a type of bacteria that causes staph infection
- *Streptococcus* – a type of bacteria that causes some human diseases

Anaerobes (*oxygen-hating*):

Preferred Environmental Conditions

- Many anaerobes do not function efficiently with oxygen present in the immediate environment
- Anaerobes may use gases other than oxygen (e.g., sulfur dioxide, carbon dioxide) for respiration

Preferred Environments

- Mammalian intestinal tracts
- Ocean and lake sediments
- Hydrothermal vents and some underground waters

Examples of Anaerobic Microorganisms

- *Clostridium* – found in all of the example preferred environments listed above for anaerobes and can cause food spoilage
- *Methanogens* – type of microorganism found in anoxic conditions such as animal digestive tracts or ocean/lake sediments that produces methane

Extremophile Data Table

NAME	PREFERS	AVOIDS	ENVIRONMENT	ADDITIONAL INFO	EXAMPLES OF ORGANISMS
Thermophiles	High temperatures 42°C to 113°C (107°F to 236°F)				
Psychrophiles					
Acidophiles					
Alkalophiles					
Barophiles					Microorganisms around deep-sea black smokers
Xerophiles					
Halophiles					
Anaerobes			Animal digestive tracts, ocean/lake sediments		

Extreme Environments Beyond Earth

For Comparison:

Earth atmosphere: 78% nitrogen, 21% oxygen, 1% argon, and 0.03% carbon dioxide, variable water content
Earth average temperature: 15°C (59°F)

Mars
- very thin and unbreathable atmosphere of 0.005 bar (1 bar is the approximate pressure on Earth at sea level)
- atmosphere contains 95% carbon dioxide, 3% nitrogen, 1.5% argon, trace amounts of water
- average temperature of -53°C (-63.4°F); temperature occasionally rises above the freezing point in equatorial regions
- polar ice caps made of CO_2 ice, water ice, and dust
- evidence for abundant surface water in the past and current possibility of abundant subsurface water

Venus
- extremely high atmospheric pressure of 90 bars (equivalent to the pressure at an ocean depth of nearly 0.6 mile)
- atmosphere contains 96.5% carbon dioxide, 3.5% nitrogen
- virtually no water at all
- clouds of sulfuric acid droplets
- mean surface temperature of 470°C (878°F)

Europa (moon of Jupiter)
- extremely thin atmosphere, with oxygen as a major constituent
- water ice surface
- strong evidence of water oceans beneath thick water ice crust
- water ice crust subject to partial melting due to tidal heating
- surface temperature of -150°C (-238°F)

Titan (moon of Saturn)
- thick atmosphere of 1.5 bars is ~ 90% nitrogen and smaller amounts of methane, ethane, and argon
- surface temperature of -180°C (-292°F)
- composed of rock and ice (ice is primarily a combination of water, methane, and ammonia)
- surface oceans likely a mixture of ethane and methane
- possible deep underground oceans of ammonia and water

Io (moon of Jupiter)
- average is about -143°C (-225°F); however, frequent lava flows cause a great range in surface temperature
- extremely active volcanism
- thin, sulfur dioxide atmosphere
- boiling sulfur and sulfur dioxide geysers with temps as high as 326°C (620°F) with measured vent temperatures as high as 2,000°C (3,632°F)

8
LIVING A POLAR LIFESTYLE: THE IMPORTANCE OF WATER FOR LIFE

GOALS
- Explore hydrogen bonding in water
- Understand how hydrogen bonding affects the density, solubility, and melting and boiling points of water
- Reason about specific heat in relation to planetary temperatures

As we explore the origins of life on Earth, we gain a better idea of the requirements for the development of life as we know it. One substance that all scientists agree on as a necessity for life on Earth is at least an intermittent presence of liquid water. In fact, water is considered so important that some astrobiologists equate the search for life in the universe with the search for water. Why is water so important? In this activity you will explore several of the characteristics of water that make it such a unique, and essential, substance on our planet.

PART A: HYDROGEN BONDING

When we think of water, we usually think of the colorless, odorless liquid that comes out of our tap, flows in our rivers and oceans, and falls from the sky in the form of rain or snow. However, to understand the properties that make it unique, one must think of water in a much different way. In Figure 1 below, we have included a drawing to provide a representation of a single water molecule. Water is a molecule that consists of one large oxygen atom (represented by the dark circle in Figure 1) and two smaller hydrogen atoms (represented by the white circles in Figure 1). This arrangement of atoms is commonly written as H_2O. It is the interactions of the hydrogen and oxygen atoms between water molecules that are the key to understanding the uniqueness of water.

When oxygen and hydrogen come together to form a bond, they share their electrons. This type of bond is called a covalent bond and is represented by the line connecting the oxygen atom to the hydrogen atom in Figure 1. However, they do not share the electrons equally. The oxygen atom will have a higher density of electrons surrounding its nucleus than the nucleus of the hydrogen atom will. This will cause the oxygen atom to be slightly more negative and the hydrogen atom to be slightly more positive. Due to this unequal distribution of electric charge, we refer to the covalent bond between the hydrogen and oxygen atom as being *polar*. Since each hydrogen atom is bonded to the single oxygen atom, there are actually two polar bonds within the single water molecule. The small arrows above the two covalent bonds in Figure 1 are used to represent the polarity of each individual bond. Because of the way the electrons within the water molecule interact, the oxygen and hydrogen atoms do not all lie in a straight line. Consequently, the polarity of each of the individual bonds adds up to make the entire water molecule itself polar, as indicated by the large arrow.

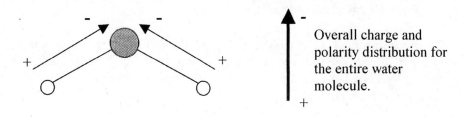

Figure 1

Activity 8

Atoms or molecules that have a net charge or that have regions with a net charge interact with each other in very specific ways. When any two objects (or regions) with a net opposite charge (i.e., one positive and one negative) are close to one another, they experience an attractive force between them. Conversely, when two objects (or regions) with the same net charge (i.e., both positive or both negative) are close to one another, they experience a repulsive force between them. Furthermore, these repulsive and attractive forces become greater when the distance between particles is decreased or if the amount of charge of either of the particles is increased.

1. Which would require less energy, moving two particles of the same charge close together, or moving two particles of opposite charge close together? Why?

2. Consider the three different arrangements of water molecules (A, B, and C) shown in Figure 2 below. Which of the following (A, B, or C) would take the most energy to assemble (or is the least stable arrangement)? Rank the arrangements in order from the greatest amount of energy to the least amount of energy required to assemble each arrangement. Explain how you decided on this ranking.

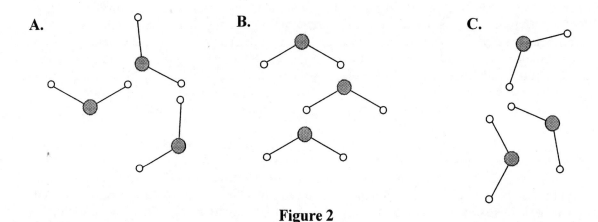

Figure 2

The electrical attraction between the slightly negative oxygen atom of one water molecule and the slightly positive hydrogen atoms of another water molecule is only one example of *hydrogen bonding*. In this type of bonding, electrons are not exchanged or shared between the atoms. Instead, the attraction between oppositely charged atoms of each molecule acts like a bond. This interaction between molecules is extremely weak compared to other types of intramolecular bonds (e.g., covalent or ionic). The relative weakness of hydrogen bonds means that the molecules must be in fairly close proximity to experience the attraction of hydrogen bonding.

3. Which of the arrangements of water molecules in Figure 2 exhibits hydrogen bonding? Specify which atoms are involved in hydrogen bonding by drawing a dotted line connecting the pair of atoms, indicating the bond between them. (Note: There may be hydrogen bonds in more than one of the arrangements A, B, or C.)

4. Which arrangement of water molecules (A, B, or C) has the greatest number of hydrogen bonds? How does this relate to the amount of energy necessary to construct that arrangement of molecules (see your responses to Questions 2 and 3)?

5. Would the arrangement you identified in question 4 be considered a stable or unstable arrangement? Why or why not?

6. Based on your responses to Questions 2 and 4, into which arrangements (A, B, and/or C) do you think freely moving water molecules will most likely arrange themselves? Explain your reasoning.

Hydrogen bonds are relatively weak when compared to covalent or ionic bonds. Therefore, hydrogen bonds between molecules can easily be broken. As a result of this weakness, hydrogen bonds are continually being formed and broken between water molecules.

The greater amount of movement (kinetic energy) each water molecule has, the easier it is to break hydrogen bonds. As temperature increases, the number of molecules engaged in hydrogen bonding decreases, such as when a liquid becomes a gas. Conversely, as temperature decreases the number of molecules involved in hydrogen bonding increases, such as when a liquid becomes a solid.

7. On average, which molecules have more kinetic energy, those that are in a gas or those that are in a liquid?

8. Based on the information in the previous paragraph, why is there less hydrogen bonding in a gas than a liquid?

9. Why is there more hydrogen bonding in a solid than a liquid?

10. If water molecules always existed in a highly structured arrangement as in arrangement B in Figure 2, do you think that water would be in a solid, a liquid, or a gas state? Why?

Activity 8

PART B: MORE DENSE THAN ICE

On average, as the temperature of water decreases, the water molecules have less kinetic energy and move much more slowly. Eventually, each molecule loses so much energy that it can no longer break the hydrogen bonds it forms with neighboring water molecules easily. Consequently, the number of molecules participating in hydrogen bonding increases. Hydrogen bonds hold the water molecules in place and create a stable structure called a *lattice*. Figure 3 illustrates an example of the lattice structure of solid water (ice).

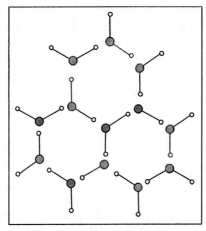

Figure 3   **Figure 4**

1. Compare Figure 3 with Figure 4. Each box is the same size, but one exemplifies water as a liquid (Figure 4) and the other represents water as a solid (ice, Figure 3). Which box contains more water molecules?

2. *Density* can be defined as the amount of particles (mass) per unit of volume. Imagine that Figures 3 and 4 actually represent cubes containing an equal volume of water. According to the definition of density provided, which is more dense, liquid water or solid water (ice)?

Examine Figure 5 on the following page. The two bottles are identical and were initially filled with the same volume of liquid water and then sealed. However, the bottle on the right was then placed in a freezer. A microscopic view of the contents of each bottle is shown in Figure 5. Imagine that each square represents an equal volume of water from each bottle.

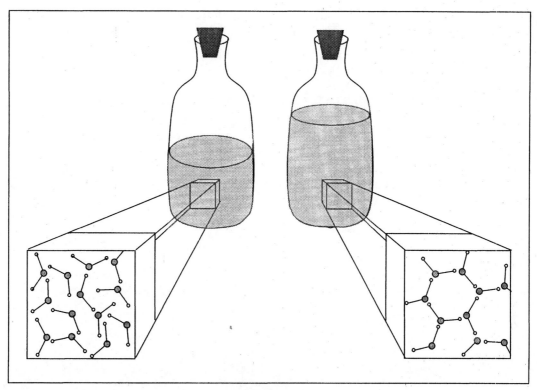

Figure 5

3. Suppose that during class you hear two students discussing the bottles in Figure 5. Carefully read the discussion below and answer the following question.

   Student #1: The level of the water in the bottle on the right is higher than the one on the left. So I think that there are more water molecules in the bottle on the right.

   Student #2: I disagree. The water level is higher in the right bottle, but it's not because there are more molecules. There is more space between the molecules so they take up more space overall.

   Do you agree or disagree with either or both of the students? Explain your reasoning for each.

4. Based on your responses to Questions 1 through 3, why do you think ice floats in water?

Activity 8

5. Suppose that instead of floating, ice sank to the bottom of oceans, lakes, and so on. What would happen to bodies of water over very long periods of extreme cold?

6. If ice did not float in liquid water, how would this affect life in oceans, lakes, and so forth?

7. Europa, one of Jupiter's moons, is known to have a surface covered by water ice. Do you think it is possible for liquid water to exist beneath the frozen surface of Europa? Explain your reasoning in detail.

8. Do you think it is possible that life exists on the surface of Europa? Why or why not?

9. Do you think it is possible for life to exist beneath the surface of Europa? Why or why not?

PART C: WATER – UNIVERSAL SOLVENT OF LIFE?

When a substance, like sugar, is dissolved in a liquid, like water, the result is a solution. The substance being dissolved is called a *solute* and the liquid in which it is dissolved is called a *solvent*. Water has the ability to dissolve more substances than any other solvent, and consequently is known as a nearly universal solvent. As a result, water is essential to many of the chemical and physical processes associated with life as we know it. In fact, life on Earth is composed primarily of water.

There are several types of particles that will dissolve (are soluble) in water. These include, but are not limited to, positively and negatively charged atoms, polar molecules similar to water, alcohols, proteins, and phospholipids. The reason why so many types of particles are soluble is due to the polar nature of water described in Part A. We will now explore how the polarity of water allows it to function as an effective solvent.

1. In the previous section we used the following symbol to represent a single water molecule. Indicate which part of the water molecule is partially positive (+) and which is partially negative (-) by drawing the symbols in the appropriate locations.

2. In Figure 6 below you will see a variety of molecules with their electric charges identified. Correctly draw how eight water molecules might arrange themselves with respect to the polar molecules in the box. For each of the polar molecules below, the regions of excess positive (+) or negative (-) charge have been identified for you. Also keep in mind that like charges repel and opposite charges attract. Two water molecules have been drawn for you, with hydrogen bonding indicated by dashed lines.

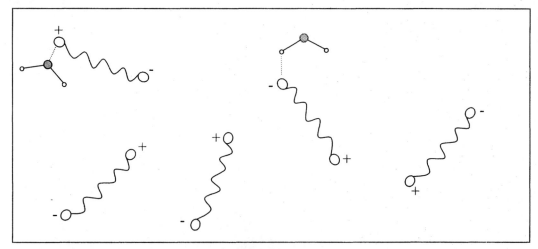

Figure 6

3. Now imagine that instead of polar molecules, the box above was filled with the same amount of nonpolar molecules. For the purposes of this activity we will make the assumption that these nonpolar molecules do not have any regions with a net charge distribution. Which would be stronger, the interaction between two polar molecules or the interaction between a polar molecule and a nonpolar molecule such as described above? Why?

4. When placed into the box with the nonpolar molecules described above (such as hydrocarbons), would the water molecules experience any hydrogen bonding with the nonpolar molecules? With each other? Explain your reasoning.

Activity 8

5. Based on your responses to the previous two questions, quickly sketch how water and nonpolar molecules might arrange themselves in the box provided below. Use the following symbol to represent a nonpolar molecule: ∿∿

Figure 7

6. Which of your two drawings (Figure 6 or Figure 7) represent a situation in which molecules are soluble in water? Explain your reasoning.

7. If water were truly a universal solvent (could dissolve everything it comes in contact with), would you be able to contain it within

    a. a Styrofoam cup? Why or why not?

    b. a plastic cup? Why or why not?

    c. a metal cup? Why or why not?

8. Life as we know it is composed primarily of water. All cells that make up living organisms are encased in a membrane composed of nonpolar components. What would be the implications for life if water were truly a universal solvent?

PART D: LIQUIDS ON EARTH

As was noted in Parts A through C, water is an excellent solvent and is necessary for many of life's functions. However, we have not considered the properties of other solvents. In the following questions we will consider the boiling and melting points of water and compare it to other solvents.

Living a Polar Lifestyle

1. What is the approximate average daily temperature (in °C) during the summer where you live? In the winter? By how many degrees does this temperature fluctuate throughout each of the seasons on average?

    Summer Average: _____   Estimated Daily Fluctuation: _____

    Winter Average: _____   Estimated Daily Fluctuation: _____

2. Water remains liquid over a large temperature range (from 0°C to 100°C, 32°F to 212°F). How do the ranges you gave in Question 7 compare with the temperature range in which water remains liquid?

3. Calculate the difference in temperature between the boiling and freezing points of each solvent below. Enter these values into the last column. One has been done for you. (Note: Room temperature is approximately 25°C.)

Solvent	Freezing Point	Boiling Point	Range in Temperature
Water	0°C	100°C	
Ethane	-172°C	-89°C	83°C
Methanol	-93.9°C	64.9°C	
Ammonia (NH_3)	-77.7°C	-33.3°C	
Hydrogen Sulfide (H_2S)	-85.5°C	-60.7°C	
Acetic Acid (Vinegar)	16.6°C	117.9°C	
Hexadecane (Similar to Gasoline)	19°C	288°C	
Phenol	43°C	181.7°C	

4. The table above lists several different solvents, including water, and the temperature ranges over which each solvent remains liquid.

    a. Which solvent has the largest temperature range?

    b. Which has the smallest temperature range?

    c. On Earth, the average range in temperature on any given day may range from just above 0°C to upward of 20°C. Which of the solvents will be a liquid over the entire temperature range on Earth?

    d. Would the solvent from the Question 4c be a good solvent for life? Why or why not?

    e. Would the solvent with the greatest range in temperature be useful as a solvent for life? Why or why not?

Activity 8

    f.   Would the solvent with the smallest range in temperature be useful as a solvent for life? Explain your reasoning.

5. Living cells as we know them on Earth are composed primarily of water. Imagine that they were composed primarily of ethane instead. Would life be able to exist over Earth's temperature range if cells were composed primarily of ethane? Explain why or why not.

6. Now imagine that living cells were composed primarily of phenol instead. Would life be able to exist over Earth's temperature range if cells were composed primarily of phenol? Explain why or why not.

7. Is it more important that a solvent has a large temperature difference between its boiling and freezing point or that this temperature difference occurs within Earth's temperature range, or are both factors important? Explain your reasoning.

8. Based on your responses to Questions 4 through 7, on which of the planets listed below would you most likely begin a search for life? Explain your reasoning.

PLANET	AVERAGE TEMPERATURES	TEMPERATURE RANGE
Mercury	- - -	-170°C to 350°C
Venus	456°C	127°C to 467°C
Mars	-53°C	-133°C to 5°C
Pluto	-226°C	-235°C to -210°C

Life in the Universe – Activities Manual, 2nd Edition   Prather, Offerdahl, and Slater

PART E: STABILIZING TEMPERATURES

We have already discussed that in order to increase the average temperature of water, energy must be added to increase the kinetic energy of the water molecules. The amount of energy required to raise the temperature of 1 gram (g) of water 1°C is 4.186 joules (J). This is known as the *specific heat* of water. Every material has a unique specific heat that depends on its chemical composition. We have listed the specific heat of other solvents in the table below.

SOLVENT	SPECIFIC HEAT [J/(g*°C)]
Water	4.186
Ethanol	2.400
Mercury	1.380
Acetic Acid (Vinegar)	2.130
Ammonia (NH_3)	0.470

1. How much energy would you need to apply to 1 g of water to change its temperature from 10°C to 11°C?

2. How much energy is required to raise the temperature of 1 g of mercury from 77°C to 78°C?

3. How much energy is required to raise the temperature 1 g of mercury from 65°C to 70°C?

4. Acetic acid has a lower specific heat than water. What does this say about the amount of energy required to change the temperature of 1 g of acetic acid as compared to the amount of energy required to change the temperature of 1 g of water by the same amount?

5. Suppose you had a substance with a specific heat greater than water. What does this say about the amount of energy required to change the temperature of 1 g of this substance as compared to the amount of energy required to change the temperature of 1 g of water by the same amount?

6. The majority of Earth's surface is covered with water. Furthermore, Earth on average experiences a temperature change of only about 20°C between day and night. Would the daily change in temperature on Earth be smaller or larger if, instead of water, Earth's surface were covered with a substance with a smaller specific heat? Explain your reasoning.

Activity 8

7. Would having oceans made of a liquid with a vastly smaller specific heat influence the ability for life to exist on Earth? Explain your reasoning.

8. Some astrobiologists equate the search for life in the universe with the search for water. Based on what you have learned throughout the entire activity, summarize all the characteristics of water that make it essential for life on Earth.

To Terraform or Not to Terraform Mars, That Is the Question

Goals
- Discuss hypothetical scenarios that could potentially impact the existence of life on Earth
- Explore possible missions to terraform the surface of Mars
- Consider societal implications of terraforming

Part A: Scenario Review

Prior to starting this activity, each member of your group should read the sheets found at the end of this activity: *Background Information, Scenarios, Mission Statements,* and *Mars Facts*.

1. What is the projected rise in sea level that would result from the West Antarctic Ice Sheet breaking free from its anchors and sliding into the ocean?

2. What is believed to be causing the increased rate at which the West Antarctic Ice Sheet is moving?

3. What percentage of Earth's population lives in the zone that would be most directly affected by the events described in Scenario A?

4. Which scenario (or scenarios) has the potential to cause a dramatic change in Earth's temperature? Explain the change in temperature that would occur.

5. Which event is more likely to inhibit sunlight from reaching the surface of Earth?

When considering the repercussions of the two scenarios, it is important to consider both the likelihood that the events described will occur on a relevant time scale and the potential danger these events present to the future of life on Earth.

6. What size was the object that is thought to have caused the last mass extinction event on Earth?

7. How much of all life on Earth is thought to have been eliminated as a result of this impact?

8. How long has it been since the last mass extinction event occurred on Earth?

Activity 9

9. What is thought to be the average time between these mass extinction events on Earth?

10. Based on your answers to the previous questions, could we experience another mass extinction event within your lifetime? What is the shortest time period (from today's date) we might expect to go by before the next one of these events occurs on Earth?

11. If Scenario A were to occur, does it have the potential to eliminate all *human* existence on Earth? If so, how would this happen? If not, why not?

12. If Scenario A were to occur, does it have the potential to eliminate all *forms of life* on Earth (including all plants, insects, and bacteria)? If so, how would this happen? If not, why not?

13. If Scenario B were to occur, does it have the potential to eliminate all *human* existence on Earth? If so, how would this happen? If not, why not?

14. If Scenario B were to occur, does it have the potential to eliminate all *forms of life* on Earth (including all plants, insects, and bacteria)? If so, how would this happen? If not, why not?

15. How long has it been since *all* forms of life were completely eliminated from Earth?

16. Which of the two scenarios presents a more immediate and inevitable threat to present-day life on Earth? Explain your reasoning.

PART B: MISSION REVIEW

1. Which mission (or missions) is focused on creating an environment on Mars that will be fit for human habitation?

2. Which mission (or missions) is focused on establishing at least the presence of some form of Earth-based life on Mars?

3. What would be the results of completing Step 1 of *Mission I*?

4. Will humans be in existence long enough to realistically consider inhabiting Mars after implementation of *Mission I*? Explain your reasoning.

5. To achieve their objectives, all of the missions rely on the Martian environment behaving as predicted and the successful development and implementation of advanced technologies. What are the most significant malfunctions, miscalculations, or incorrect assumptions that could cause a catastrophic outcome for each mission?

    Mission I:

    Mission II:

    Mission III:

6. Mission II and Mission III will each require that we have a source of energy available for use on the surface of Mars. What energy source could we use on Mars to meet these energy requirements?

Activity 9

7. Why is the production of ozone and CO_2 an essential component of each mission?

PART C: MISSION COMPARISONS

1. Based on the perceived immediacy and possible threat to life on Earth presented in Scenario A, discuss which mission should be pursued. Which factors are most important to consider (e.g., time, resources, money, potential threat to life, etc.)? Rank the missions from most important to pursue to least important to pursue. Explain the reasoning behind your ranking.

2. Based on the perceived immediacy and inevitability of Scenario B, discuss which mission should be pursued. Which factors are most important to consider (e.g., time, resources, money, potential threat to life, etc.)? Rank the missions from most important to pursue to least important to pursue. Explain the reasoning behind your ranking.

PART D: SOCIAL IMPLICATIONS

1. Are there any moral or ethical reasons why any one of the missions might be less appealing than the other? Why or why not?

2. Do we, as humans, have an obligation to prolong the human species by transferring a portion of our population to Mars, thus mitigating the likelihood of a singular extinction event? Explain your reasoning.

3. What can we learn from the terraforming process that can help us improve or take care of Earth?

4. Throughout Earth's history, grand-scale environmental changes (i.e., changes in climate, global weather, planetary surface changes) have occurred over millions of years. With human intervention, grand-scale changes on Mars could happen over a significantly smaller time interval and with a much greater magnitude. What are the negative or potentially catastrophic results that could occur from our attempts to terraform Mars?

5. Suppose that a population of microorganisms already exists in the subsurface of Mars. What are the risks for preexisting Mars life if life from Earth is introduced into the Martian environment?

6. Consider the expenses of a mission to terraform Mars. Who should pay for the mission and where should the funds come from?

7. What form of government, if any, should be established on a successfully terraformed Mars? Who from Earth should be represented?

Background Information

Imagine yourself stepping off an airplane and feeling almost light enough to fly. The sky is pale blue and there is not a cloud as far as you can see. The air is thin and brisk and remarkably fresh and invigorating. The horizon is interrupted by little more than scant vegetation, a few insects and birds, and a thin layer of dust being shifted by the cool wind. Between you and the horizon is a flat expanse of land intermittent with rolling hills and a few jagged escarpments of rock, all of which is colored a unique shade of orange. The Sun, in its bright, yellow brilliance, is directly above you and though it is warm on your skin, it appears more distant than you remember. As your gaze shifts toward the terminal, you see a few hundred strangers with warm smiles that appear to be eagerly anticipating your approach.

You have arrived at your destination. Your flight was not a 5-hour nonstop. Your view out the passenger window was not of towering mountains, flowing rivers, endless oceans of blue, or of lights that anonymously dot the nighttime countryside. Your plane was not a 747 streaming across the upper atmosphere. And your destination was not a small, cozy island in the Caribbean. Your flight was a harrowing 6-month journey. Your view out the passenger window was that of an endless expanse of black punctuated by distant pinpoints of light. Your plane was an interplanetary shuttle drifting through the frigid vacuum of space. Your destination was the red planet-Mars.

Sound implausible? Well it might be beyond our reach now, but this may change in the future. Mars has long been considered a likely place for the existence of life. Since the early 1800s, scientists have suspected the presence of life on Mars. In 1895, Percival Lowell was convinced that Martians had constructed a vast system of canals to transport water from the frigid polar ice caps to the tropical latitudes for the purpose of irrigating crops. In 1938, thousands of people panicked, thinking that Martians were invading Earth as Orson Wells aired *War of the Worlds* on public radio. In recent times, the concept of little green men running around on Earth has been dropped. Through vastly improved technology and observation techniques, Mars has shown to be, apparently, lifeless. By way of that same technology, scientists have been speculating about the possibilities of transplanting life to Mars from Earth.

Scientists have for some time been aware of factors that can limit the lifetime of a planet's habitable environment. It is now becoming common knowledge that human actions could be rapidly diminishing the lifetime of complex life (humans, plants, animals) here on Earth. While the notion of exploration of new frontiers has existed since long before Columbus landed on the "New World," necessity may be adding increased impetus toward taking such a step again. NASA and other space exploration organizations around the globe have been investigating the possibilities of not only finding life on other planets, but actually placing it there. Physical manipulation of the surface and atmosphere would certainly be required and Mars is the most plausible destination for such an endeavor to take place. While Mars is the planet in our solar system most like Earth, it would require chemical, mechanical, physical, and quite possibly biological engineering achievements on a magnitude that have not yet been accomplished on Earth. All of this would be necessary if humans were to someday inhabit the surface of Mars.

Although Mars is similar to Earth, in many ways it is also unique. Mars has a 24.5-hour day, much like Earth, but because of its distance from the Sun, a Martian year is 687 Earth days long. Similar to Earth, Mars has ice caps at its poles that advance and recede with the changing of the seasons. On Mars, however, the ice caps are composed primarily of frozen CO_2. Our current evidence points to a water-rich history on the surface of Mars. There are numerous channels that appear to have been made by the erosive movement of water. In addition, there is evidence that water was often released by the heat of meteor impacts. While the surface of Mars is covered primarily by

Background Information Continued

basaltic rock and dust, there is evidence that carbonate soils also exist, further substantiating the existence of surface water in the past. In fact, some scientists speculate that there was an ocean on Mars that averaged 10 meters in depth. Furthermore, a mineral called hematite was recently discovered on Mars. Hematite is a mineral that is found on Earth in the vicinity of volcanic regions like Yosemite National Park. The existence of this mineral on Mars suggests that hot water may have been available on the surface of Mars over a time period that could have been long enough for life to have formed.

We do not know whether life exists on Mars now or has in the past. Whether or not life may be sustainable there in the future, we cannot say for sure. Extensive study of Mars may be the only way to find out. The missions described here are designed to not only search for past life on Mars, but to establish life and to propagate a habitable environment for humans on Mars in the future.

Scenarios

There have been many events in Earth's history that have dramatically impacted the nature of life on this planet. The following two scenarios describe two potential life-altering events that could occur during our time on Earth.

SCENARIO A: GLOBAL WARMING

I. History

- 120,000 years ago Earth was very similar to today (i.e., temperature, sea level, CO_2 levels, etc.).
- Research shows that there was an instantaneous 20-foot rise in sea level. This abrupt rise in sea level triggered an ice age lasting 100,000 years. These changes dramatically altered the climate and nature of life on Earth.

II. Present Day

- 90% of the ice on Earth is concentrated in Antarctica.
- 13.5% of the planet's ice is concentrated in and around the West Antarctic Ice Sheet which lies anchored to the west side of the Trans-Antarctic Mountains.
- The highest recorded warming on Earth has taken place in Antarctica within the last 50 years. This unprecedented rise has caused the average temperature in Antarctica to increase by 4.5°F.
- The West Antarctic Ice Sheet is moving 25% faster than normal due to this dramatic increase in temperature.
- If this glacier were to continue sliding toward the sea at its present rate, it could eventually break free from its anchors and slip into the ocean, causing an instantaneous 20-foot rise in sea level.
- Approximately 50% of the human population lives within 20 feet of sea level.
- A massive reduction in ocean temperatures would take place affecting many forms of sea life and therefore affect the rest of Earth's food chain.
- Because the ice would take many years to melt, it would float out into the main bodies of ocean and could reflect as much as 4% of the solar energy that presently hits the surface of Earth.
- A 4% reduction in energy absorbed by Earth's surface from the Sun could catapult us into the next ice age.

Scenarios Continued

SCENARIO B: ASTEROIDS

I. History

- Asteroid impacts have caused several mass extinction events in Earth's past.
- On at least four occasions life has been completely eradicated by asteroid impacts.
- Evidence suggests that impacts have occurred that produced enough energy to evaporate the oceans, liquefy Earth's crust, and send ejected material into the atmosphere and outer space.
- Impacts of this magnitude would have raised the temperature on Earth to nearly 2000°C (far above the temperature where any life could exist).
- The most recent of these "Earth-sterilizing" events likely occurred about 3.8 billion years ago. It is important to note that it is only from that point in time until now that life has been continuously present on Earth.

II. Implications of Impacts

- Impacts by asteroids as big as 10 km across are calculated to occur every 30 to 100 million years.
- 65 million years ago (mya) an asteroid approximately 10 km wide impacted Earth resulting in the death of 70% of all life on Earth, including the extinction of the dinosaurs. Results of this impact include:
    a. a crater on the surface of Earth nearly 200 km wide.
    b. the production of enough energy to temporarily increase the temperature of the oceans, resulting in the death of a significant amount of aquatic life.
    c. the production of a cloud of dust obscuring sunlight for hundreds of years and causing a dramatic decrease in the average global temperature of Earth.
- Craters currently exist on the Moon that are 100 times as large, over 1,000 km across.

Mission Statements

If we wish to ensure that humans continue to exist, we may one day need to actively pursue the possibility of establishing life on another planet. The three mission statements below describe potential plans that we may employ to achieve this. If the environment of Mars is going to be transformed so that it is suitable for life, we will need to significantly increase the pressure of the atmosphere. In particular, we must increase the concentration of UV-protecting ozone and CO_2 in order to help promote an active greenhouse effect. These changes to the atmosphere will help raise the average temperature on the surface and allow for the presence of liquid water on the Martian surface.

MISSION I: MICROBIAL

Objective: Seed the surface of Mars with hardy microbial life-forms that will generate a pressurized atmosphere rich with greenhouse gases through natural processes in order to create a planetary environment suitable for the evolution of life.

Time Scale: 50,000 to 500,000 years

Cost: 10 times the U.S. national debt of approximately $2 trillion

Plan of Action:

1. Utilize global windstorms to introduce anaerobic "rock-eating" bacteria (chemolithotropes) to Mars. Dust storms will serve a dual purpose: efficiently spreading bacteria around the planet and shielding bacteria from harmful UV rays. Chemolithotropes break down carbonate soils releasing CO_2 into the atmosphere. This step will be repeated until a marked increase in atmospheric pressure has been noted. Increases in the atmospheric pressure will consequently create higher temperatures.

2. Release photosynthetic bacteria (phototrophs) utilizing the windstorm process. Phototrophs employ a photosynthetic process from which oxygen is released into the atmosphere. Oxygen, upon reaching the upper atmosphere, is dissociated by the Sun's rays and recombines creating ozone. Ozone in the upper atmosphere creates a more permanent shield against UV radiation. This step will be repeated until the level of UV reaching the Martian surface has been diminished sufficiently to allow for plant life.

3. When temperatures reach sufficient levels to melt ice at lower latitudes, then various plant species will be introduced. All plants must be either self-pollinating or wind-pollinating because the carbon dioxide-rich atmosphere at this point is still too toxic to allow insects to pollinate. All such plant species will have been genetically adapted under conditions that mimic those on an evolving Mars. Microbial and plant species are expected to play a key role in the regulation of nitrogen, carbon, and mineral cycles planetwide.

4. Allow sufficient time for the temperature and atmosphere to stabilize and hope that life will continue to evolve toward more complex life-forms.

Mission Statements Continued

MISSION II: MECHANISTIC

Objective: To engineer an environment on the surface of Mars that would be suitable not only for bacterial and plant life but also for human habitation in the near future.

Time Scale: 5,000 to 50,000 years

Cost: 1,000 times the U.S. national debt of approximately $2 trillion

Plan of Action:

1. Utilize planetary orbiters to "sprinkle" dark dust on the polar CO_2 ice caps to reduce planetary albedo and increase absorption of solar energy.

2. Employ orbiting mirrors to concentrate the Sun's rays on the polar caps and other regions of the planet.

    (Note: Steps 1 and 2 are designed to increase global temperatures, which causes the release of carbon dioxide, an effective greenhouse gas. Excessive heating to specific parts of the planet could theoretically cause increases in global wind and dust storms, which could help in heating the Martian surface.)

3. Create and globally distribute specialized nano-robots (nano-robots are computerized or mechanized devices that possess biological components and properties) to break down carbonate soils and release carbon dioxide.

4. Once a pressurized atmosphere has been created, construct power plants and pumps to transport potential reserves of subsurface water to liberate oxygen through surface evaporation. This allows oxygen to move into the upper atmosphere for the purpose of creating ozone. Ozone is necessary for shielding pending life-forms from UV radiation.

5. Construct and maintain genetic engineering labs and greenhouses for the purpose of manipulating organisms to live and reproduce under the conditions of the Martian atmosphere.

6. When desired temperatures, gas levels, and UV radiation levels have been met, various forms of biota could be introduced on the surface to assist in the transition and pave the way for human habitation.

Mission Statements

MISSION III: HUMAN, MACHINE, AND MICROBE

Objective: To utilize every means available in order to create not just a habitable, but a hospitable environment on Mars that resembles that on Earth.

Time Scale: 500 to 5,000 years

Cost: Unlimited financial commitment and additionally the potential for a substantial loss of human life

Plan of Action:

1. Conduct an initial investigation of the entire Martian surface to establish locations of past or present water reserves.

2. Construct several "bases" around the Martian surface from which exploration and research can take place. In particular bases near the poles where water may be available and several at "equatorial" latitudes should be established.

3. Carry out extensive searches for existing life before implementing seeding process.

4. Distribute dark dust on the polar CO_2 ice caps to reduce planetary albedo and increase absorption of solar energy. Employ reflective orbital mirrors to concentrate solar energy on the polar caps for the purpose of liberating carbon dioxide. An increase in carbon dioxide will mean an increase in atmospheric pressure.

5. Utilize nano-robots and anaerobic "rock-eating" bacteria (chemolithotrophs) to break down carbonate soils to release carbon dioxide into the atmosphere. Use global windstorms to spread bacteria.

6. Establish an efficient process for getting underground reserves of liquid water to the surface where it is to be stored in an artificially warmed and pressurized environment, but one in which the Sun's rays can reach. Water is to be utilized for human consumption and irrigation.

7. Photosynthetic bacteria will be introduced to the Martian surface as well as into the stored water. This will allow oxygen to accumulate in the atmosphere for the purpose of photo-dissociation to create ozone.

8. Greenhouses will be constructed for the purpose of engineering plants suitable for growth on the Martian surface. Crops will be grown for human consumption.

9. Mining and industry centers will be established during this process to utilize materials found on or near the surface for the purpose of reducing construction costs.

10. When a safe atmospheric pressure and a habitable temperature have been realized, further human migration will take place.

Mars Facts

Due to its proximity and physical similarities to Earth, Mars has presented itself as a likely candidate for terraforming. Below we have provided a list of useful physical characteristics of Mars.

Mars Facts

- Rotation period is 24.5 hours.
- Orbital period is 687 days.
- Distance from the Sun is 227,900,000 km (1.52 AU).
- Diameter is 6,787 km (Earth's diameter is 12,756 km.).
- Average temperature is -53°C (-63.4°F).
- Mass is 6.58×10^{23} kg (0.11 times the mass of Earth).
- The atmospheric pressure of Mars is 7,000 times less than the atmospheric pressure of Earth at sea level. This extremely low pressure makes it impossible for liquid water to remain stable on the surface of Mars.
- Mars's atmosphere consists of 95% CO_2 with trace levels of N_2.
- The surface of Mars receives an excess of 100 times the UV radiation of Earth due to the lack of ozone or other UV-absorbing layer.
- Mars's rust-colored surface is due to oxidized iron minerals.
- Global storms are frequent on Mars, with winds often in excess of 200 mph.

10
INTERSTELLAR REAL ESTATE: DEFINING THE HABITABLE ZONE

GOALS
- Define the "zone of habitability"
- Understand the range of the habitable zone within our solar system
- Explore the impact of temperature and spectral class on habitable zones
- Analyze the habitable zones for newly discovered planetary systems
- Consider the relationship between planetary mass and ability to maintain an appreciable atmosphere

PART A: WHY IS LIFE ABUNDANT ON EARTH?

Examine the information provided in the table below and answer the following questions. Note that $0.8 M_{Earth}$ means that the planet has a mass that is 80% the mass of Earth.

PLANET CHARACTERISTIC	VENUS	EARTH	MARS
Planet Mass (M_{Earth})	$0.8 M_{Earth}$	$1 M_{Earth}$	$0.1 M_{Earth}$
Planet Radius (R_{Earth})	$0.95 R_{Earth}$	$1 R_{Earth}$	$0.5 R_{Earth}$
Distance from Sun (D_{Earth})	$0.7 D_{Earth}$	$1 D_{Earth}$	$1.5 D_{Earth}$
Average Surface Temperature	456°C	10°C	-95°C
Atmosphere	Thick	Medium	Very thin

1. Which of the characteristics listed in the table above most likely have the greatest influence on whether or not life can flourish on Earth but not on Venus and Mars? Explain your reasoning.

Activity 10

2. Describe how any of the planet characteristics would change if the following were to occur.

   a. Earth was located closer to the Sun.

   b. Earth was located farther from the Sun.

   c. The Sun's temperature and size were much greater.

   d. The Sun's temperature and size were much smaller.

The planets in our solar system orbit the Sun at different distances. Scientists have developed a system for describing distances in our solar system based on the average distance between Earth and the Sun. The astronomical unit (AU) is the average Sun-Earth distance, approximately 149,570,000 km. Distances between objects in our solar system are measured using the AU as the common unit of distance. The table below provides each planet's name and average orbital distance to the Sun.

Interstellar Real Estate

3. Convert the distances from km to AU for each of the planets in our solar system. The distance for Earth is already provided for you.

Planet Name	Dist. to Sun in Km	Dist. to Sun in AU
Mercury	57,950,000 km	
Venus	108,110,000 km	
Earth	149,570,000 km	1 AU
Mars	227,840,000 km	
Jupiter	778,140,000 km	
Saturn	1,427,000,000 km	
Uranus	2,870,300,000 km	
Neptune	4,499,900,000 km	
Pluto*	5,913,000,000 km	

Pluto may or may not be defined as a planet at this moment.

The presence of liquid water at the surface of a planet appears to be one of the central characteristics distinguishing whether or not a planet can harbor life. This requires that the planet be a distance from the central star where the temperature is neither so low that all the water will freeze nor so high that all the water will boil. The region around a star where the temperature is "just right" for liquid water to be present is known as the *zone of habitability*. For a star like our Sun, a reasonable estimate for the zone of habitability would be between approximately 0.84 AU and 1.7 AU.

4. Based on the calculations in Question 3, which planet(s) would be included in the zone of habitability for our solar system?

5. Which of the planets in our solar system have the potential for liquid water on the surface? Explain how you can tell.

Life in the Universe – Activities Manual, 2nd Edition — Prather, Offerdahl, and Slater

Activity 10

6. Is our Moon in the zone of habitability? Does the Moon have liquid water on the surface? Explain your reasoning.

7. Describe how the location of the zone of habitability would change if the central star's temperature were to increase. Consider both the inner and outer boundaries of the zone of habitability in your description.

PART B: WHAT INFLUENCES THE SIZE OF THE HABITABLE ZONE?

You will now utilize the set of *Star Cards* found at the end of the activity. Examine each card and, using the information provided, sort the stars according to their distance from Earth, sequencing them from the greatest distance to the least distance.

1. Does a star's temperature appear to depend on its distance from Earth?

2. Now rearrange the *Star Cards* by temperature, sequencing them from coolest to hottest. Consider the following three characteristics: temperature, distance, and spectral class. Which characteristics most strongly influence the size and location of the habitable zone? Explain your reasoning for each.

3. In our search for life in the universe, which of the stars described on the *Star Cards* would you explore first? What spectral classes did you choose? Explain the reasoning behind your choices.

Scientists are currently searching for life around stars most similar to our Sun, those in the F, G, and K classes. Size, temperature, and brightness are important factors in identifying these stars.

4. How would the zone of habitability be different around an "F" star as compared to our Sun? Explain your reasoning.

5. How would the zone of habitability be different around a "K" star as compared to our Sun? Explain your reasoning.

Up to this point we have identified that the *just right* condition for life is the presence of liquid water on a planet's surface. This suggests we should first search for a planet that rests within the zone of habitability of a Sun-like star (classes F, G, or K.) In addition to appropriate distance from a star, a planet should also have a suitable atmosphere. Planets that are too small may have too thin of an atmosphere, and larger planets may have an atmosphere too thick to support life as we know it. For example, a sizeable atmosphere is necessary to maintain a stable temperature range on the planet, neither too hot nor too cold. However, an impenetrable atmosphere may prohibit a useful energy source, sunlight, from reaching the surface. In addition to the thickness, atmospheric composition is also important to consider in the search for life on other planets. To maintain a suitable atmosphere we will assume in this activity that a planet should have a mass between 0.5 and 10 Earth masses, with a radius between 0.8 and 2.2 times that of Earth.

STAR (TEMP./CLASS)	PLANET NAME	DISTANCE IN AU	MASS	SIZE
Altair (7,900K / A)	Jomikal	0.05	72.4 M_{Earth}	10.7 R_{Earth}
Regulus (11,500K / B)	Korbin	0.75	1.5 M_{Earth}	1.1 R_{Earth}
Procyon (6,600K / F)	Heino	0.55	7.7 M_{Earth}	3.5 R_{Earth}
Beta Cassiopeia (8,000K / F)	Atkeins	1.25	8 M_{Earth}	2.1 R_{Earth}
Alpha Centauri (5,750K / G)	Kauli	0.46	250 M_{Earth}	175 R_{Earth}
Epsilon Indi (4,400K / K)	Aardal	1.6	0.3 M_{Earth}	0.42 R_{Earth}
Epsilon Eridanus (4,600K / K)	Roski	1.5	0.9 M_{Earth}	1.75 R_{Earth}
Barnard's Star (2,700K / M)	Lackeve	2.2	195 M_{Earth}	182 R_{Earth}

Activity 10

6. Which of the hypothetical planets listed in the table above would be strong candidates for life? Explain why or why not for each planet.

Jomikal: Kauli:

Korbin: Aardal:

Heino: Roski:

Atkeins: Lackeve:

PART C: OTHER PLANETS IN HABITABLE ZONES?

Refer to the page listing *Recently Discovered Planets* at the end of this activity. The table shows actual data corresponding to planets outside our solar system orbiting other stars.

The "primary" star is classified by spectral type. The mass of the planet orbiting each star is provided in multiples of Earth's mass. The distance from the planet to its companion star is shown in AU.

For example the table on the *Recently Discovered Planets* sheet shows that the star named HD 16874 is a G type star. The orbiting planet has a mass of about 68 times the mass of Earth. We also find that the planet is located 0.35 AU from its star.

1. Examine the planets in the table. In the space below, list the star name and state whether or not the companion planet is likely to support life. To assist you in making this determination use the *Planet Classification Flowchart* provided at the end of the activity. For planets that fail the test, state the major factors that keep them from being strong candidates for life.

2. Could any of the planets you listed have an orbiting moon that has the ability to support life? List any possible candidates and explain your reasoning for each.

Now read through Scenario A and Scenario B below. Each hypothetical scenario describes a planet and moon orbiting a star. For each scenario, think carefully about the variables that impact the ability of a planet to support life.

Scenario A

Planet Dajin orbits an F type star (with a temperature of 6900K) at an average distance of 1.2 AU. Dajin has a mass of 40 M_{Earth} and a radius of 70 R_{Earth}. A moon called Thalakos orbits Dajin at an average distance of 0.15 AU with a period of 1 week. The mass of Thalakos is 2.0 M_{Earth} and its radius is 0.85 R_{Earth}.

Scenario B

Planet Abjure orbits a G type star (with a temperature of 5700K) at an average distance of 0.5 AU. Abjure is 8.0 M_{Earth} and a radius of 2.0 R_{Earth}. Orbiting this planet is a moon called Sengir that orbits Abjure once every 6 months. Sengir orbits Abjure at an average distance of 0.05 AU and has a mass of 1.2 M_{Earth} and a radius of 1.0 R_{Earth}.

3. Is it possible that life could exist on the planet described in Scenario A? Explain how it's possible or, if not, what characteristics inhibit the existence of life on this planet?

Activity 10

4. Is it possible that life could exist on the moon described in Scenario A? Explain how it's possible or, if not, what characteristics inhibit the existence of life on this moon?

5. Is it possible that life could exist on the planet described in Scenario B? Explain how it's possible or, if not, what characteristics inhibit the existence of life on this planet?

6. Is it possible that life could exist on the moon described in Scenario B? Explain how it's possible or, if not, what characteristics inhibit the existence of life on this moon?

Star Cards

Star Name: Vega

Temperature: 9,900 K
Distance: 26.3 Light-Years
Spectral Class: A

Star Name: Sirius

Temperature: 9,600 K
Distance: 8.6 Light-Years
Spectral Class: A

Star Name: Spica

Temperature: 25,500 K
Distance: 262 Light-Years
Spectral Class: B

Star Name: Regulus

Temperature: 13,260 K
Distance: 78 Light-Years
Spectral Class: B

Star Cards

Star Name: Altair

Temperature: 8,000 K
Distance: 16.8 Light-Years
Class: A

Star Name: Procyon

Temperature: 6,600 K
Distance: 11.4 Light-Years
Spectral Class: F

Star Name: Sun

Temperature: 5,800 K
Distance: 0 Light-Years
Spectral Class: G

Star Name: AlphaCentauri

Temperature: 5,750 K
Distance: 4.3 Light-Years
Spectral Class: G

Life in the Universe – Activities Manual, 2nd Edition   Prather, Offerdahl, and Slater

Star Cards

Star Name: Epsilon Indi

Temperature: 4,400 K
Distance: 11.8 Light-Years
Spectral Class: K

Star Name: Epsilon Eridanus

Temperature: 4,100 K
Distance: 10.8 Light-Years
Spectral Class: K

Star Name: Barnard's Star

Temperature: 2,800 K
Distance: 6 Light-Years
Spectral Class: M

Star Name: Kapteyn's Star

Temperature: 3,500 K
Distance: 12.8 Light-Years
Spectral Class: M

Star Cards

Star Name: Beta Cassiopeia

Temperature: 7,000 K
Distance: 46 Light-Years
Spectral Class: F

Star Name: Lambda Cepheus

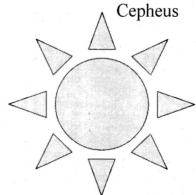

Temperature: 33,000 K
Distance: 49 Light-Years
Spectral Class: O

Star Name: Delta Orion

Temperature: 30,000 K
Distance: 815 Light-Years
Spectral Class: O

Recently Discovered Planets

Use the *Planet Classification Flowchart* to decide which, if any, of these planets or their moons might be capable of supporting life similar to that found on Earth. To the left of each entry, tell if you would explore each planet for life or not, and why.

STAR	SPECTRAL CLASS	DISTANCE IN AU	MASS
HD83443	K 0	0.038	111 M_{Earth}
HD83443	K 0	0.174	51 M_{Earth}
HD16874	G 5	0.35	68 M_{Earth}
HD10814	F 8	0.98	108 M_{Earth}
Epsilon	K 2	3.3	273 M_{Earth}
Gliese	M 4	0.13	41 M_{Earth}

Life in the Universe – Activities Manual, 2nd Edition — Prather, Offerdahl, and Slater

Planet Classification Flowchart

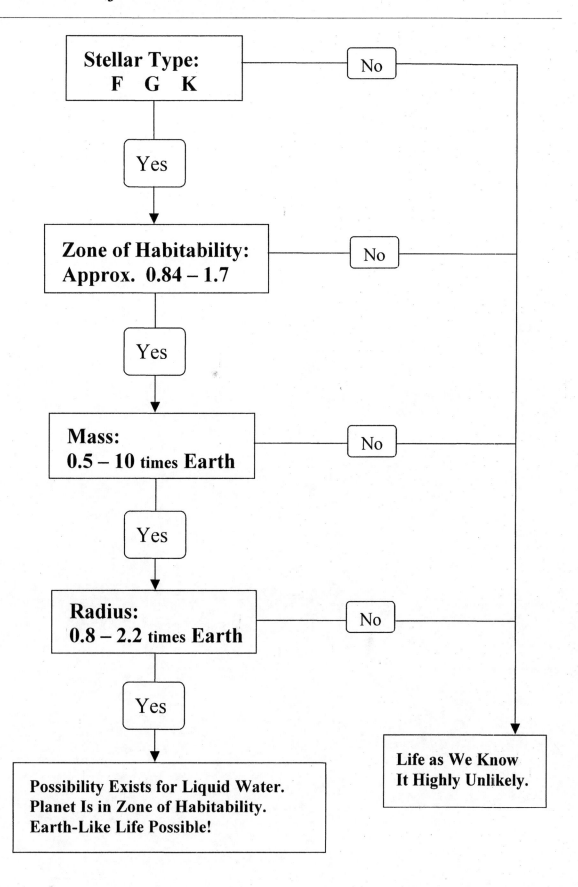

11
WOBBLING STARS: HOW EXTRASOLAR PLANETS ARE DISCOVERED

GOALS
- Discover how Doppler shift is used to detect the presence of extrasolar planets
- Analyze graphical data to interpret extrasolar planet motion
- Compare the orbital distance and mass of extrasolar planets

PART A: THE MOTION OF THE SUN

Astronomers have made astounding progress in discovering planets orbiting stars outside our solar system. In fact, they have identified a vastly larger number of these planets, called extrasolar planets, than currently exist in our own solar system. There is more than one technique for detecting extrasolar planets. But with current technology, the most effective technique for detection has been the radial velocity of Doppler shift technique. In this activity, you will learn how astronomers use this technique to infer the presence of planets around other stars.

Let's begin by looking at the radial velocity technique applied to our own solar system. In Figures 1 and 2, there are two different depictions of our solar system. Since the Sun and Jupiter account for nearly all the mass of our solar system, our solar system is modeled here as a two-body problem involving only the Sun and Jupiter. Note that these representations are not drawn to the proper scale for the size or distance of the objects shown.*

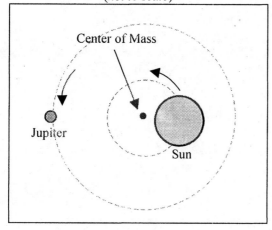

Figure 1

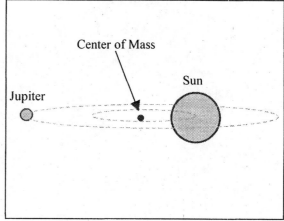

Figure 2

*The center of mass of the solar system is located within the Sun. We've exaggerated the Sun's orbit about the center of mass.

Activity 11

1. Is Jupiter coming toward or going away from you in Figure 2?

2. Is the Sun coming toward or going away from you in Figure 2?

3. Draw a stick figure in Figure 1 to indicate where an observer would need to be in relationship to the solar system to see the view shown in Figure 2.

Now examine the four drawings in Figure 3 below. Each of the four drawings shows the positions of the Sun and Jupiter at a different time during a single orbit.

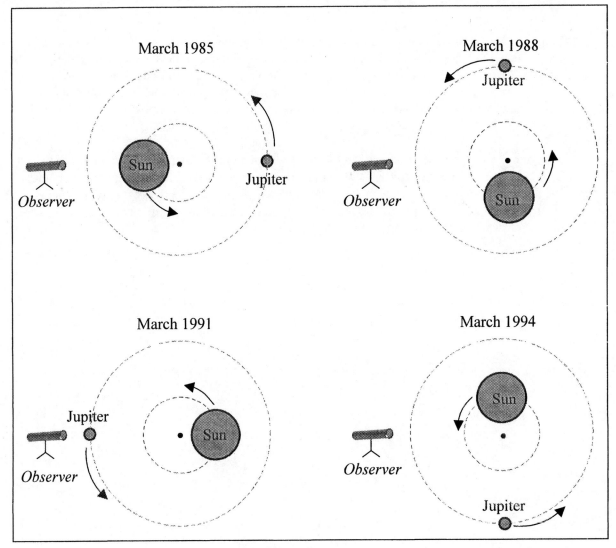

Figure 3

4. In Figure 3, does the Sun always appear to remain in the same position? If not, describe its motion.

Extrasolar Planets

5. What form of interaction or force causes the orbital motions of the Sun and Jupiter?

6. Estimate the time (in Earth years) for the Sun to complete one orbit (this time is known as the orbital period). How does this time compare to the orbital period of Jupiter?

7. For each of the four drawings in Figure 3, use the boxes below to draw what the observer would see at each time period if he or she was observing the solar system edge-on. See the example in Figure 2.

March 1985	March 1988
March 1991	March 1994

8. Make two sketches below (using representations in Figures 1 and 2) depicting what you would see in September 1992 from the observer location. Your drawings need to include the positions of the Sun and Jupiter.

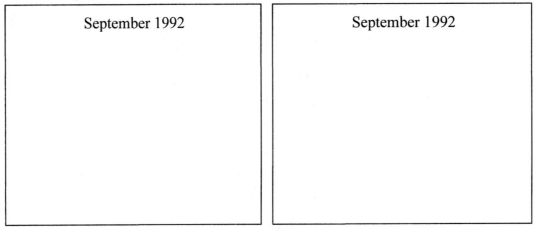

September 1992	September 1992

When studying motion it is useful to consider the object's velocity as being made of two parts or components. The component of velocity that is directed toward (negative) or away from (positive) the observer's line of sight is known as the *radial* velocity.

9. Imagine that you are at the observer location shown in the drawings you made in Question 8 for September 1992, and that you are located much farther away from the Sun and Jupiter's orbital paths than is depicted in your drawing.

   a. From your point of view and line of sight at the observer location, would the Sun appear to be moving with a radial velocity? If so, is it positive or negative? Explain your reasoning.

   b. From your point of view and line of sight at the observer location, does Jupiter appear to be moving with a radial velocity? If so, is it positive or negative? Explain your reasoning.

10. Now consider the entire interval shown in Figure 3 from March 1985 all the way through March 1994.

    a. During which range of dates would the Sun appear to be moving with a radial velocity? When is the radial velocity positive and when is the radial velocity negative?

    b. During which range of dates would Jupiter appear to be moving with a radial velocity? When is the radial velocity positive and when is the radial velocity negative?

If an observer and a star being studied are both stationary then the wavelength of the light traveling from a star to an observer will remain unchanged. However, if the star is moving toward an observer (with a negative radial velocity), then the light's wavelength will appear to be shifted to a shorter wavelength (or blue shifted). Furthermore, if a star is moving away from an observer (with a positive radial velocity), the light's wavelength will appear to be shifted to a longer wavelength (or red shifted). This shifting in the wavelength of light due to the motion toward or away from a light source (like a star) is known as the Doppler shift.

11. During which range of dates would the light from the Sun have a Doppler shift to a longer wavelength? Explain your reasoning.

12. During which range of dates would the light from the Sun have a Doppler shift to a shorter wavelength? Explain your reasoning.

13. If, instead of viewing the solar system edge-on (like in Figure 2), an observer was very far away from the solar system and looking directly down on the solar system, during what time interval, if ever, would the observer see the Sun have a Doppler shift to shorter wavelengths? Explain your reasoning.

PART B: DISCOVERING NEW PLANETS

Astronomers measure the change in radial velocity using the Doppler shift of the light coming from a star. They can graph this change in radial velocity versus time. Figure 4 shows a radial velocity versus time graph for a star that has an extrasolar planet orbiting it.

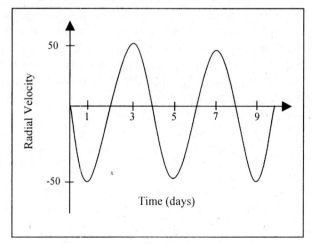

Figure 4

Activity 11

1. At what time(s) is the star moving with a radial velocity that is zero?

2. Imagine you are observing this star edge-on (like in Figure 2 from Part A). Sketch how it would look when the radial velocity is zero.

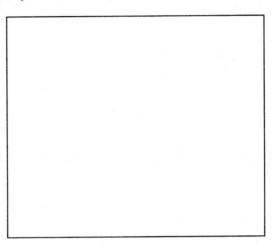

3. At what time(s) would the star be moving *toward* the observer with the greatest radial velocity? How fast would it be moving?

4. At what time(s) would the star be moving *away from* the observer with the greatest radial velocity? How fast would it be moving?

5. What do astronomers observe about the light of stars that makes it possible to determine the stars are moving with a changing radial velocity?

Now let's examine actual data gathered by astronomers in their pursuit to find planets orbiting distant stars outside our solar system. Below are the radial velocity versus time graphs for three stars (47 Ursae Majoris, 49 Sengir V Cdc, and HD 11964).* Two of the graphs come from measurements of stars with companion planets; the other is a graph of a star without a companion planet. Note that the dots shown in each graph represent the actual measured radial velocities for these stars, and the curves provide a "best fit" to the data points. Use these best fit curves to answer the questions regarding the motion of these three stars.

---

* The original versions of the real graphs can be found at http://cannon.sfsu.edu/~gmarcy/planetsearch/doppler.html .

47 Ursae Majoris

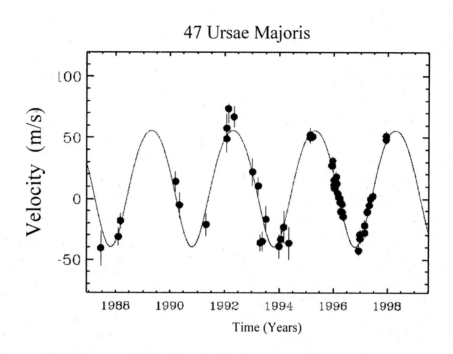

Sengir V Cdc

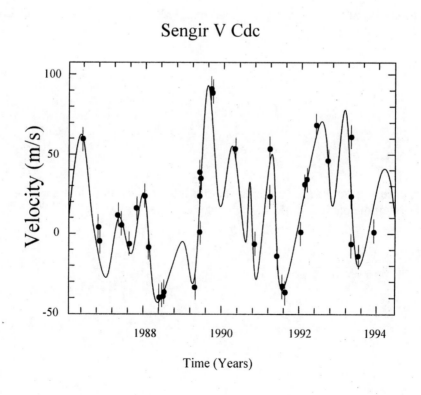

HD 11964

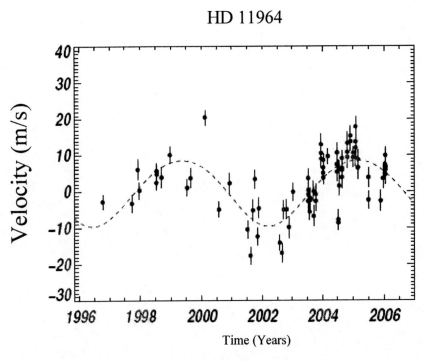

Time (Years)

6. At what time(s) was each star measured to be moving toward Earth with the greatest radial velocity? Note that you are to use a point on the best fit curve in each graph and not the individual data points. How fast was the star moving?

    47 Ursae Majoris   Sengir V Cdc   HD 11964

7. For each star, state whether or not you think the star has a companion planet and, if so, estimate the orbital period of the planet. If not, explain why not.

    47 Ursae Majoris   Sengir V Cdc   HD 11964

PART C: EXPLORING SYSTEM PROPERTIES

As was mentioned in Part A, the number of extrasolar planets discovered to date is far greater than the total number of planets within our solar system. However, we have not yet discussed the nature of these extrasolar planets. How close do they orbit their parent star? What are their orbital periods? How massive are they? In this part of the activity, you will look at real data of extrasolar planets discovered using the radial velocity technique showing how these extrasolar planets compare to planets within our solar system.

Figure 5 shows a histogram for 167 extrasolar planets. A histogram is a type of graph that shows information sorted into bins. Notice that the last bin on the far right represents the number of extrasolar planets with a mass 15 times that of Jupiter. The height of the bar in each bin represents how many extrasolar planets have that particular planetary mass. It is important to note that all of the planet masses are represented in units of Jupiter mass, M_{Jup}, so that the mass of the planet in the last bin on the right would be written 15 M_{Jup}.

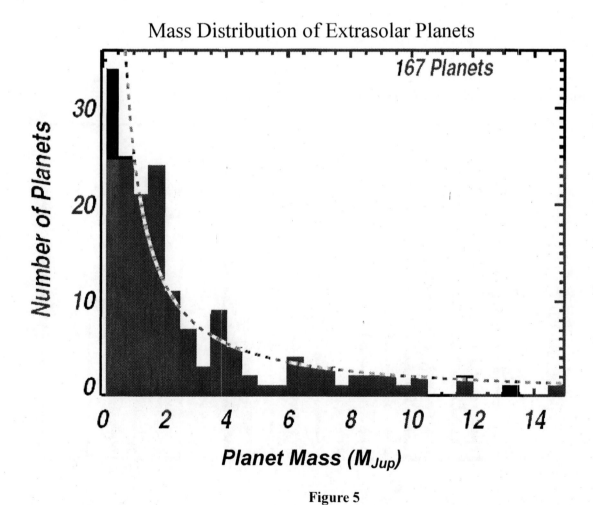

Figure 5

1. What is the mass of the planets, in units of Jupiter mass, indicated by that bar corresponding to the number 9?

Activity 11

2. If Jupiter is approximately 300 times more massive than Earth, how many times greater than the mass of Earth are the smallest planets on the chart in Figure 5?

3. You overhear two students in class having the following conversation:

   Student #1: I don't know why people say that it will be difficult to find Earth-size planets orbiting around other stars. I mean, if you look at the bar on the far left in Figure 5, you can clearly see that there are at least 30 planets with a mass about the same as Earth.

   Student #2: I think the graph shows that we are finding planets that are way bigger than Earth. The masses shown in the graph are based on the mass of Jupiter, not Earth. So even the bar that is less at a value less than 1 will be hundreds of times more massive than Earth.

   Do you agree or disagree with one, both, or neither of the students? Why?

Figure 6 is another histogram, this time showing the number of planets that orbit at a particular distance from their central star (as determined by their semimajor axis).

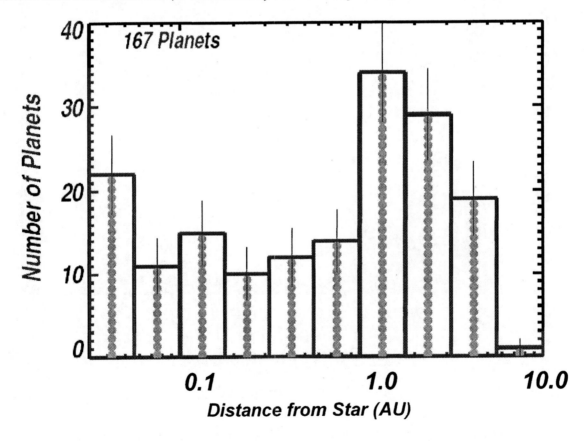

Figure 6

Extrasolar Planets

4. Approximately how many extrasolar planets orbit at the same distance that Earth orbits the Sun (1 AU)?

5. Jupiter orbits our Sun at a distance of 5 AU. Are the majority of extrasolar planets orbiting closer to their stars or farther from their stars than Jupiter orbits the Sun? Explain why you think so.

6. A family friend calls you and says she has read the most astounding headline in the newspaper. The headline reads, "*Much Like Our Solar System, Scientists Are Finding Jupiter-Sized Extrasolar Planets at Very Large Distances from Their Star.*" What would you correct about this headline for your friend?

Figure 7 is a histogram showing the number of extrasolar planets that have a particular orbital period. Note that the orbital periods of these planets are recorded in units of days.

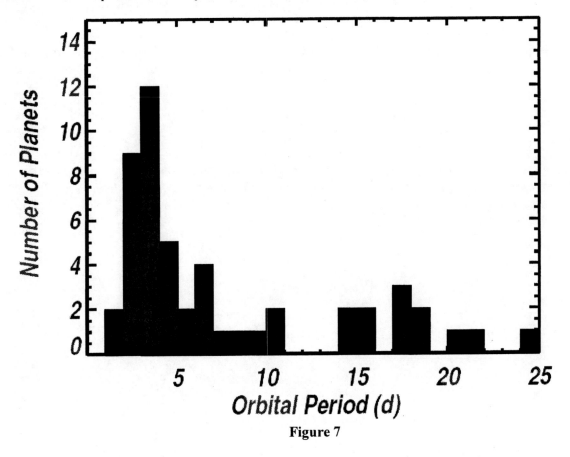

Figure 7

7. How long will it take a planet represented by a 10 on the *X*-axis of the histogram above to complete one orbit of its central star?

8. Approximately how many times more quickly would the planet in Question 7 orbit its star as compared to Earth (*circle one*)?

        3 times faster

        30 times faster

        300 times faster

        3,000 times faster

9. All of the extrasolar planets shown in Figures 5 through 7 above were detected using the radial velocity technique. In just a few sentences, describe how these extrasolar planets compare in general to Jupiter in terms of mass, orbital distance, and orbital period.

10. Imagine you and your classmates set out to detect the next extrasolar planet using the radial velocity technique. Which of the following planets would you most likely discover? Why?

	Planet A	Planet B	Planet C
Mass	2 M_{Jup}	7.5 M_{Jup}	11 M_{Jup}
Distance from Star	1 AU	10 AU	0.5 AU
Orbital Period	3 days	12 days	23 days

11. What do the distributions in Figures 5, 6, and 7 above tell us about the likelihood that we will find Earth-like planets with the radial velocity technique? Explain your reasoning.

12
THE RARE EARTH:
HOW RARE IS EARTH-LIKE LIFE?

GOALS
- Interpret graphs relating luminosity, stellar mass, and main-sequence lifetime
- Investigate the time frame for complex life to develop on Earth
- Reason about how stellar abundance can be used to approximate the rarity of Earth-like planets with complex life in our galaxy

PART A: HOW OLD, HOW LUMINOUS, AND HOW MASSIVE SHOULD A STAR BE TO SUPPORT COMPLEX LIFE?

The *Rare Earth hypothesis* suggests that Earth-like planets containing complex (multicellular) life as we know it are probably quite rare in the Milky Way Galaxy. Scientists generally agree that Earth formed about 4.5 billion years ago, yet complex life has existed on Earth for only about the last 600 million years. It is still unclear exactly what events led up to the emergence of complex life on this planet. A particularly important condition that scientists believe to be necessary for the development of complex life is a long period of relatively stable climate resulting from a steady planetary orbit at the correct distance from an appropriate type of star.

Let's begin our search for an appropriate star by looking at the characteristics of our Sun that make it possible for complex life to flourish on Earth. During the activity, the symbol ☉ will be used to identify characteristics of our Sun. The Sun is a G-type star in the *main-sequence* phase of its life. This means that it is engaged in the stable conversion of hydrogen into helium by nuclear fusion in its core. The Sun radiates energy mostly in the form of visible light. The measure of a star's radiation energy is called its *luminosity*. Although it has changed somewhat, the Sun has been emitting approximately the same amount of radiation for about 5 billion years, making it about halfway through its entire *main sequence lifetime* of about 10 billion years. We will investigate the portion of the Rare Earth hypothesis that considers how the lifetime, mass, and luminosity of a star can influence the possibility of complex life developing on an orbiting planet. You will also calculate the fraction of stars in the Milky Way Galaxy that fall within a particular stellar mass range, which could allow for the development of complex life.

Examine Graphs 1 and 2 near the end of this activity. Graph 1 shows how the main sequence lifetime of a star is related to the star's mass. Graph 2 shows how the star's luminosity is related to its mass. Refer to these graphs to answer the following questions. Note that on both graphs, the axes for mass are exactly the same.

1. Describe how the main sequence lifetime changes as stellar mass increases (Graph 1).

2. Describe how stellar luminosity changes as stellar mass increases (Graph 2).

3. On Graph 1, first mark the position on the stellar mass axis that corresponds to the Sun's mass. Then mark the position on the main sequence lifetime axis that corresponds to the Sun's lifetime. Place a ☉ mark for the Sun on the graph where these values intersect.

Activity 12

4. On Graph 2, first mark the position on the stellar luminosity axis that corresponds to the Sun. Then mark the position on the stellar luminosity axis that corresponds to the Sun. Place a ☉ mark on the graph where these values intersect.

For Questions 5 and 6, refer to BOTH Graphs 1 and 2.

5. Do stars live for a longer or shorter period of time, and are they brighter or dimmer, when they have a stellar mass that is *less than* the Sun's?

6. Do stars live for a longer or shorter period of time, and are they brighter or dimmer, when they have a stellar mass that is *greater than* the Sun's?

7. Refer back to the introductory paragraph. How long did it take before complex life developed on Earth? What fraction of the Sun's entire main sequence lifetime is this?

8. Based on your answer to the previous question, estimate the minimum amount of time that a main sequence star can exist and still have a planet that has time to develop Earth-like complex life. Explain why you think so.

9. Use the time you estimated in Question 8 as the minimum amount of time that a main sequence star needs to exist and still have a planet that has time to develop Earth-like life. Mark this value on the main sequence lifetime axis on Graph 1. Label this point t_{min}. Find the stellar mass that corresponds to t_{min}. Label it on the stellar mass axis as either M_{min} or M_{max}. How did you decide which should it be? (Note: It may be useful to use a ruler or straight edge to place your marks.)

10. On Graph 2, mark the stellar mass that you identified in question 9 on the stellar mass axis. Now find the stellar luminosity that corresponds to this mass. Label it on the stellar luminosity axis as either L_{min} or L_{max}. How did you decide which should it be? (Note: It may be useful to use a ruler or straightedge to place your marks.)

11. On Graph 2, place a mark on the stellar luminosity axis for a star that is *brighter* than the star you marked in Question 10. Find the stellar mass that corresponds to this luminosity. Label this stellar mass Star 1.

12. Again on Graph 2, place a mark on the stellar luminosity axis for a star that is *dimmer* than the star you marked in Question 10. Find the position on the main sequence lifetime axis that corresponds to this star and label it Star 2.

13. On Graph 1, clearly label the points on the stellar mass axis that correspond to Star 1 and Star 2.

14. On Graph 1, locate the stellar lifetime associated with Star 1. Is it possible for complex life to exist on a planet around Star 1? Why or why not?

15. Locate the stellar lifetime associated with Star 2. Is it possible for complex life to exist on a planet around Star 2? Why or why not?

16. Based on your answers to Questions 14 and 15, was the mark you labeled L_{min} or L_{max} in Question 10 the correct limit? (If you wrote L_{min}, does it represent the minimum stellar luminosity for complex life? If you wrote L_{max}, does it represent the maximum stellar luminosity for complex life?) Explain your reasoning.

PART B: HOW BRIGHT IS TOO BRIGHT FOR LIFE?

The hotter and more luminous a star is, the more radiation it gives off, and most of this radiation tends to be at shorter wavelengths. The Sun emits most of its radiation in the visible wavelength range. In general, the shorter the wavelength of the emitted radiation, the more damaging it is to complex life because shorter wavelengths are higher energy than longer wavelengths. For Earth-like life-forms, ultraviolet radiation is highly damaging to cells and to DNA. In Graph 3, wavelength of the stellar radiation is plotted on the vertical axis, and stellar luminosity is plotted on the horizontal axis. Use this information to answer the following questions.

1. Look at the stellar radiation wavelength axis of Graph 3. Which direction (going up the vertical axis or going down) represents an increase in energy?

Activity 12

2. Now find the location of visible radiation on the stellar radiation wavelength axis and determine the stellar luminosity that corresponds with this frequency. How does this value compare to the Sun's luminosity?

3. Place a mark on the stellar luminosity axis that represents your estimate for the limit on how bright a star can be before the wavelength of most of its radiation is too damaging for complex life to develop easily. Label this mark as either L_{min} or L_{max}. How did you decide which it should be?

4. In Part A you made an estimate for L_{max} and marked it on Graph 2. In the previous question, you made an estimate for L_{max} and marked it on Graph 3. Although the scales on these two graphs are different, do both your estimates of L_{max} predict the same outcome for the possibility of complex life existing?

5. In Part A, what variable was affecting your estimate for L_{max}? In other words, what was limiting how luminous a star can be and still have complex life on a nearby planet?

6. In Question 2 from Part B, what variable was most affecting your estimate for L_{max}? What was limiting how luminous a star can be and still have complex life on a nearby planet?

7. Could complex life exist on a planet near a star that (a) has a main sequence lifetime slightly longer than the time it took for complex life to develop on Earth, and (b) gives off radiation that is mostly in the UV range? Explain your reasoning.

8. Could complex life exist on a planet near a star that (a) has a main sequence lifetime much shorter than the time it took for complex life to develop on Earth, and (b) gives off radiation that is mostly in the visible range? Explain your reasoning.

Part C: How Many Stars Is That?

Complex life requires a minimum amount of energy to develop. For instance, we might establish that stars with a luminosity of less than $0.3\ L_\odot$ are unable to provide enough energy to support complex life on a nearby planet. Recall that the symbol $\odot$ relates to characteristics of our Sun.

1. Using the luminosity limit from the paragraph above, place a mark (L_{min}) on Graph 1 that represents the lower limit on stellar luminosity for complex life.

2. Find the stellar mass that corresponds to L_{min} and label it as either M_{min} or M_{max}. How did you decide which it should be?

3. Can stars with a stellar mass less than this have planets with complex life? Explain your reasoning.

4. Now that you have established an L_{min} and an L_{max} for Graph 1, write down the possible range of stellar masses (in units of $M_\odot$) for stars that can support complex life.

Consider the information shown in Graph 4. The horizontal axis represents the range of stellar masses for *main sequence stars*. This type of graph, in which information is sorted into bins, is called a *histogram*. Since stars with a mass less than $0.5\ M_\odot$ are quite dim there is not a large sample of data for these small mass and dim stars. As a result, Graph 4 starts with data representing stars that have a mass of $0.5\ M_\odot$ and extends to stars with a mass of $10\ M_\odot$. Each bin of data in Graph 4 represents a range of $0.1\ M_\odot$ stellar masses. Notice that the first bin on the left represents stellar mass between 0.5 and $0.6\ M_\odot$, the second bin represents stellar mass between $0.6\ M_\odot$ and $0.7\ M_\odot$, the third bin represents stellar mass between $0.7\ M_\odot$ and $0.8\ M_\odot$, and so on. The last bin on the right is stellar mass between $9.9\ M_\odot$ and $10\ M_\odot$. The height of the bar in each bin represents how many stars exist within a specific range of stellar masses. Multiply the number on the Y-axis by 100,000,000 to get the number of stars in that mass range in the galaxy. The total number of stars represented in this histogram is approximately 40,000,000,000. Although the total number of stars in the Milky Way is approximately 200 billion, the vast majority are very low mass, dim stars, which are not in the range of stars we are actively searching for extraterrestrial life and are not included in Graph 3.

5. How many stars exist with masses between $0.5\ M_\odot$ and $0.9\ M_\odot$? How many of these stars could support complex life?

Activity 12

6. According to this histogram, how many total stars in the galaxy can support complex life? Use your mass range from Question 4 of Part C.

7. What fraction of all main sequence stars can support complex life? Express your answer as a fraction, a decimal, and a percentage.

The histogram shown in Graph 4 provides an ESTIMATE of the total number of stars with main-sequence lifetimes. These main sequence stars make up approximately 70% of all stars in the galaxy. The remaining stars fall into four broad classes: those with too little mass (e.g., brown dwarfs); those with too much mass (e.g., blue giants); those that are too young (e.g., T Tauri stars); and those that are too old (e.g., red giants, white dwarfs, and neutron stars).

8. Using Graph 4, out of ALL the stars in our galaxy, how rare (what percentage) are stars that could support complex Earth-like life?

Graphs

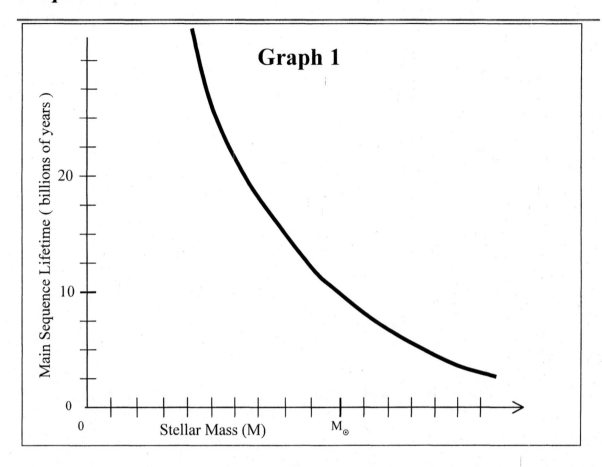

Graph 1

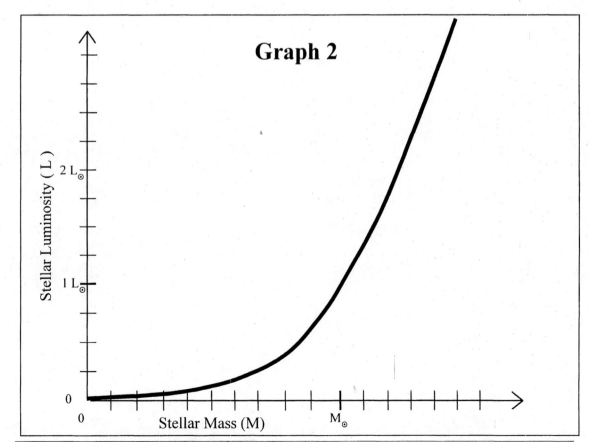

Graph 2

Graphs

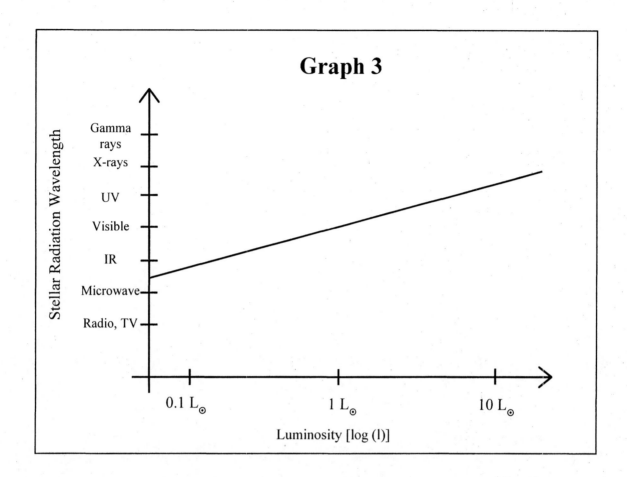

Graphs

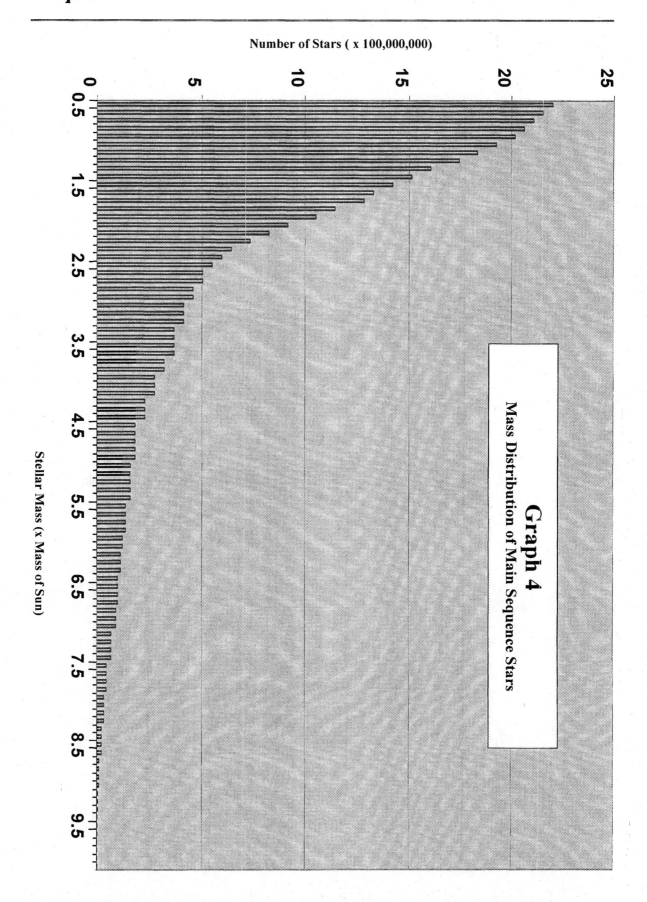

13

THE DRAKE EQUATION:
ESTIMATING THE NUMBER OF CIVILIZATIONS
IN THE MILKY WAY GALAXY

GOALS
- Develop estimation techniques
- Perform extrapolations from real data
- Examine the range and definition of each variable from the Drake Equation
- Evaluate how changes in the variables of the Drake Equation influence the outcome

PART A: HOW MANY CRATERS ARE ON THE MOON?

Below is a 2.7 square kilometer (2.7 km^2) image of the *Apollo 14* landing site on the Moon.* The Moon's surface can be divided up into 14 million such patches.

* *The original public domain image can be found at: http://www.nasm.edu/APOLLO/AS14/a14landsite.htm*

1. Estimate the number of craters on the Moon that are half the size of a football field (approximately 50 meters).

Life in the Universe – Activities Manual, 2nd Edition

Activity 13

2. Describe why your estimate might increase or decrease if a different picture of the Moon's surface were used.

3. How would your estimate change if you were estimating the number of craters that are:

   a. a quarter the size of a football field?

   b. two or more times the size of a football field?

PART B: MAKING COMPLEX ESTIMATES

There are many instances in science where estimation is much more useful and efficient than counting. In particular, estimation techniques are important when analyzing a system for which counting is not actually possible. Complete the following estimation task.

1. Predict how many female psychology majors with long hair are going to be at the library between 6:00 p.m. and 7:00 p.m. in the reference section reading a journal article by a male author.

   Prediction:_____

2. To check your prediction, complete the following table by estimating the variables listed below. The notes column is provided for you to describe the reasoning behind the estimates you will make. We have already included two example notes with possible estimates to help guide your work.

The Drake Equation

VARIABLE	ESTIMATED VALUE	NOTES
n – total number of students at your college/university		
f_f – fraction of females at your college/university		
f_p – fraction of those students that are psychology majors		A reasonable estimate might be that 1/5 of all students are psychology majors.
f_L – fraction of female students with long hair		
f_t – fraction of students at the library between 6:00 p.m. and 7:00 p.m.		
f_R – fraction of students in the reference section of library		
f_F – fraction of students in reference section reading a journal article		For the students in the reference section, a reasonable estimate might be that 1/8 of them are reading a journal article
f_m – fraction of journal articles by a male author		
$F_{f,p,L,t,R,F,m}$ – number that represents the product of all the fractions above		
T – number of female psychology majors with long hair at the library between 6:00 p.m. and 7:00 p.m. in the reference section reading a journal article by a male author (**F** x **n**)		
CLASS AVERAGE, T_{avg}		

Using the chart above, we can calculate T, the number of female psychology majors with long hair at the library between 6:00 p.m. and 7:00 p.m. in the reference section reading a journal article by a male author (**F** x **n**), using the equation:

$$T = (F_{f,p,L,t,R,F,m}) \times n = (f_f \times f_{fp} \times f_{,L} \times f_{ft} \times f_R \times f_{,F} \times f_m) \times n$$

Activity 13

3. Explain how the value of T would change if you observe males instead of females. Provide an example calculation with your written explanation.

4. Explain how the value of T would be different if you changed your definition of long hair? Provide an example calculation with your written explanation.

5. Suppose that you were making this estimate for an all-female school between 6:00 p.m. and 9:00 p.m. How would your estimate change? Why?

PART C: USING THE DRAKE EQUATION

The equation shown below is a version of the Drake equation. This equation can be used to provide a rough estimate of the number of communicating civilizations in the Milky Way Galaxy. It is important to note that there are a number of other variables that could be considered when making this estimate that we will not include in our calculations.

$$N_c = n_s \times f_p \times n_e \times f_l \times f_i \times f_c \times L$$

In Part B, you estimated the number of students that had particular characteristics. In this portion of the activity, you will use the same techniques to estimate N_c, the number of existing extraterrestrial civilizations in the Milky Way Galaxy that possess the technology to communicate beyond their home planet.

The Drake Equation

1. Complete the table below. Before making your estimations, carefully read the *Drake Equation Background Information Sheet* found at the end of this activity.

DESCRIPTION OF VARIABLE	ESTIMATED VALUE	NOTES
n_s – number of target stars in the galaxy that: • are second-generation stars with heavy elements • are hot enough to have a large habitable zone • have long enough lifetimes for life to develop	$n_s =$	
f_p – fraction of those stars with planets or planet systems	$f_p =$	
n_e – number of planets in a planetary system that are at the right temperature for liquid water to exist (in the habitable zone)	$n_e =$	
f_l – fraction of these planets where life actually develops	$f_l =$	
f_i – fraction of these planets with at least one species of intelligent life	$f_i =$	
f_c – fraction of these planets where the technology to communicate beyond their planet develops	$f_c =$	
L – lifetime of communicating civilizations divided by the age of the galaxy, 10 billion years	$L =$	
Now calculate N_c using the above estimates:		
N_c – number of communicative civilizations	$N_c =$	

$$N_c = n_s \times f_p \times n_e \times f_l \times f_i \times f_c \times L$$

2. What value did you get for N_c (the number of civilizations)?

3. How does the value change if you double the lifetime of communicating civilizations?

Life in the Universe – Activities Manual, 2nd Edition   Prather, Offerdahl, and Slater

Activity 13

4. How does the estimate change if we discover that only one third of Sun-like target stars have planets?

5. How would you change your estimate if we discovered that early life developed on both Venus and Mars?

6. Determine the most reasonable maximum and minimum values that your group believes the terms f_p, n_e, f_l, f_i, and f_c could have. Record your values for each term below.

7. Calculate the range of values for N_c that result from using the maximum and minimum values that your group recorded in the previous question.

8. Do the maximum and minimum values that you calculated make sense to your group? Explain why you think they might be too large, or too small, or just right.

9. How many intelligent, communicating species in the galaxy do we actually know about? What, then, is the actual minimum value for N_c? (Hint: It is not zero.) Explain your reasoning.

When scientists make estimates with the Drake equation, they often use the following values. If we think that all stars that are like our Sun have planets, then we could estimate $f_p = 1$ to represent 100%. If we use our solar system as a model for estimation, then there is only one planet in the habitable zone that we know has liquid water on its surface (Earth) so we could imagine setting $n_e = 1$. And, since Earth is the only planet in our solar system that we know to have developed life, it seems reasonable to set $f_l = 0.1$ to represent that approximately 1 out of every 10 planets has life. It is essentially impossible to know the fraction of species that develop on a planet that turn out to be intelligent and able to communicate. Therefore, a conservative estimate for f_i and f_c that we might use is 0.1 for each term. As a rough guess we might imagine that across the galaxy intelligent communicating civilizations last for about 20,000 years out of the 10 billion year existence of the galaxy, which sets $L = 2 \times 10^{-6}$

10. What value do you get if you use the estimates provided in the preceding paragraph? How does this value compare to your original estimate, your estimate for a maximum value, and your estimate for a minimum value?

11. Scientists have discovered many gas giant planets orbiting other stars outside our solar system. Some of these planets orbit within their star's habitable zone (where liquid water can exist). Describe how these findings could change your estimates.

Drake Equation Background Information Sheet

$$N_c = n_s \times f_p \times n_e \times f_l \times f_i \times f_c \times L$$

n_s – This number represents how many billions of stars in the galaxy meet the following two criteria:

1. The star must be a second- or third-generation star formed from an interstellar cloud that included the necessary heavy elements for life (e.g., carbon, oxygen, etc.). The elements are created during the evolution of first-generation, super-massive stars and supernova events that occurred early in the history of our galaxy. A reasonable estimate for this number is 200 billion stars.

2. The star must release enough energy to have a sizeable habitable zone. A habitable zone is the region around a star where liquid water could exist on an orbiting planet. Ninety percent of the stars in our galaxy are too cool to have a sizeable habitable zone. This eliminates stars with spectral type K5 and cooler. Of the remaining 10%, nearly a quarter of those have lifetimes too short for complex life to develop. This eliminates stars with spectral type F8 and warmer as they have lifetimes shorter than 4 billion years.

Our Sun, a G2 star, fits both of these categories and thus is one of the target stars. Such target stars are often referred to as Sun-like stars. A reasonable estimate for the number of target stars is: $2 \times 10^{11} * 10\% * 75\% = 15$ billion stars.

f_p – This number represents the fraction of those stars meeting the above criteria that also have planets or planet systems around them. Recent discoveries of numerous extrasolar planets suggest that most stars like our Sun probably have planets.

n_e – This number represents how many planets there are at the right temperature for liquid water to exist (i.e., in the habitable zone). Recent discoveries suggest that we should also consider including moons around gas giant planets that are orbiting their central star in the habitable zone. A reasonable estimate for this number is difficult to imagine. In our solar system, the number ranges from one to three depending on if you include Venus or Mars. If Saturn were to migrate into the habitable zone, its 22 moons would make this number much larger.

f_l – This number represents the fraction of these planets where life actually develops. Some scientists believe that the evolution of life is inevitable when the conditions are right. Alternatively, we know of only one instance where life has successfully developed (Earth), therefore it is difficult to estimate this fraction.

f_i – This number represents the fraction of these planets where at least one species of intelligent life evolves. Intelligent life could develop early on some planets and later on others and again, it is difficult to estimate this fraction.

f_c – This number represents the fraction of these planets where the technology to communicate beyond the planet exists. In our own civilization, we have been using television and radio signals for nearly a century. These signals have leaked into outer space and might be detectable by extraterrestrial civilizations. As before, it is extremely difficult to estimate this number.

L – This number represents the fraction of the number of years that communicating civilizations have existed out of the total lifetime that the galaxy has existed. We call this fraction of years "Lifetime." This number depends on both social issues and technological issues. It is possible that intelligent civilizations elsewhere in the galaxy have existed for millions of years and may or may not choose to communicate beyond their own planet. Alternatively, when civilizations develop the technology to communicate they might simultaneously develop technology capable of making their environment uninhabitable (e.g., weapons of mass destruction). These factors make this number extremely difficult to estimate. L could range from 1×10^{-8} (100 years/10,000,000,000 years) to 1×10^{-4} (millions of years/10,000,000,000 years) or more.

14

IS THERE ANYBODY OUT THERE?

GOALS
- Debate the elements that represent our civilization
- Consider strategies to send information into space
- Decipher and transfer digital information
- Consider the implications of broadcasting information

A fundamental part of the quest to understand our place in the universe rests on whether or not we are alone. Throughout history, many people have proposed that intelligent life is common in our galaxy of hundreds of billions of stars, whereas others believe that we are likely the only intelligent civilization in our galaxy. Both perspectives draw on philosophical perspectives and scientific arguments but lack compelling scientific evidence. One approach to searching for other intelligent life is to intentionally send information describing who we are and where we are located in the cosmos. In this activity, we'll explore what information we could send and how this information should be sent.

PART A: REPRESENTING OUR INTELLIGENT CIVILIZATION

1. One way that people have cataloged the nature of a civilization is through time capsules. A time capsule is a container that holds objects representing many facets of a civilization at a particular instant in time. As a group, create a list of 10 items that could fit into a single box (3 feet on each side) that could serve as a "snapshot" of our civilization. Choose your items based on the premise that the time capsule will be sent on a spacecraft destined for travel beyond our solar system with the possibility of encountering intelligent life. Include a reason why you chose each item.

	ITEM	REASONING
1		
2		
3		
4		
5		
6		
7		
8		
9		
10		

Activity 14

2. Compare your list to that of a nearby group. What specific aspects of human civilization did the other group try to convey with their time capsule that were different from your group's choices?

3. Are there any items the other group included that you feel you should have or vice versa? If so, what are they and why would you include them in your time capsule? What item(s) currently in your time capsule would you eliminate in order to accommodate for the item(s) from the other group?

4. Due to the size limitations of what we can send into space using our current technologies, it might be more efficient to attach a plaque to the spacecraft depicting the nature of life on Earth and different aspects of our world. What information would you engrave on a 6-inch by 9-inch metal plaque intended to travel far into outer space with the possibility of encountering other intelligent life?

5. How would you depict the information listed in Question 4 on your plaque? What pictures, words, shapes, etc. would you use? On a separate piece of paper, provide an example *sketch* of your plaque.

In the early 1970s two spacecraft, *Pioneer 10* and *11*, were launched from Earth, destined to be the first human-made objects to escape our solar system. Each of the spacecraft carried a 6-by-9-inch plaque designed to depict when the spacecraft was launched, from where it was launched, and by what kind of life. Below is a drawing of the actual plaque sent on *Pioneer 10* and *11*. Several pieces of information are represented pictorially on the plaque. Some of the information included is:

    a. The average height of a human woman in relationship to the *Pioneer* spacecraft
    b. The location from where *Pioneer* was launched
    c. The location of our solar system within the Milky Way Galaxy
    d. The distances of easily identified pulsars relative to the Sun
    e. The order of the planets within our solar system as well as the relative distance (in binary form) between the planets
    f. A gesture of good will by humankind

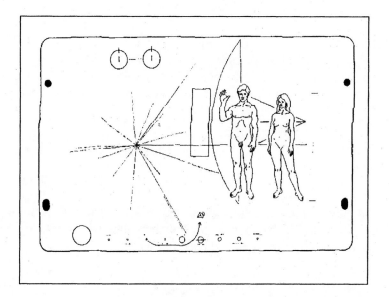

Figure 1

6. Examine the drawing of the *Pioneer 10* and *11* plaque provided in Figure 1. Identify at least four of the six pieces of information listed above that are depicted on the plaque. Describe in words how these four pieces of information were represented on the plaque.

7. Compare the information you listed in Question 4 with the information from the *Pioneer* plaque. Is there any information included on the *Pioneer* plaque that you would like to include on your plaque or vice versa? If so, what is it and why would you include it? If not, why not?

Activity 14

8. Do you feel the information depicted on the *Pioneer 10* and *11* plaques is the most important information to communicate with an extraterrestrial civilization? Is there any information about the nature of life on Earth that was not included on the *Pioneer* plaques or on your group's plaque that you didn't mention before that you feel is important to communicate? Explain your reasoning.

PART B: SELECTING A METHOD OF COMMUNICATING

1. The *Pioneer* spacecraft are traveling at enormously high speeds – more than 70,000 km/hr. At this speed, it takes them about 25,000 years to travel a distance of one light-year (a light-year is the distance light travels in one year.) How many years would it take the *Pioneer* spacecraft to reach the Alpha Centauri star system approximately 4 light-years (ly) from Earth?

2. In the science fiction movie *Contact*, radio signals were detected emanating from the bright star Vega at a distance of 26 ly from Earth. Approximately how long would it take a spacecraft like *Pioneer* to make a trip from Earth to Vega and back again?

3. Radio waves, like all forms of electromagnetic radiation (light), travel at the speed of light. How long would it take radio waves to travel to an object one light-year away?

4. How long would it take radio waves to make a trip from Earth to Vega and back again?

5. If a radio signal were sent to Vega today and it was immediately returned by an extraterrestrial civilization and received back here on Earth, how old would the youngest person in your group be when the radio signal was received back on Earth?

In Part A, your group created (1) a list of 10 items to be included in a time capsule that would serve as a current snapshot of civilization and (2) a list of necessary information about the nature of life on Earth in general to be engraved on a plaque. With this in mind, answer the following questions.

6. Would the information you included in your time capsule still be relevant if it had been sent to Vega via spacecraft and intercepted by intelligent life? Why or why not?

7. Would the type of information represented on the *Pioneer* plaques still be relevant if intercepted by intelligent life? Why or why not?

8. Does the amount of time for a message to be sent via spacecraft influence the type of information you would send (a snapshot of civilization vs. information on the nature of life on Earth)? Explain your response in detail.

PART C: SENDING A MESSAGE

As we learned in Part B, radio signals, which travel at the speed of light, are a much quicker method for sending - or receiving - information about intelligent civilizations through the immense distances between stars. Therefore, it is useful to learn how to convey information in a digital format that can be transferred as electronic information via radio waves.

Digital pictures, such as the ones you download to a computer from the Internet, are actually composed of tens of thousands of individual pieces of information, called *picture elements*, or *pixels* for short. For example, you could use 72 pixels to represent the letters *D N A* as:

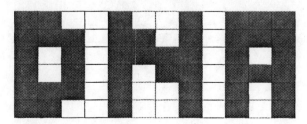

Activity 14

This could be converted into a numeric sequence of 1s and 0s, where 1 means the square is colored black and 0 means the square is colored white. Starting at the top left of the grid and moving to the right, much as one reads a book, the message would be converted as follows:

[grid image showing binary values]

or

110010010111 111011010111 101011110101 101010110111 111010010101 110010010101

1. If a "1" represented a white square and a "0" represented a black square, would this change the meaning of the message? Explain why or why not.

2. Notice that we broke up the string of 1s and 0s into groups of 12, with each group representing one row on the grid. How would the message look if we did not do this?

3. Using the 6 x 12 grid shown below, decipher the following message:

011100000100 111110001010 111110010001 001000001110 001000001010 001000001010

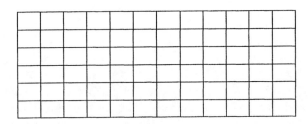

What is this? _____

4. How does the message from Question 3 differ from the first message we looked at in Part C?

5. Consider the drawing you made in Question 5 of Part A. Describe, in words, how you might represent the information from Question 5 as a digital picture sent via radio waves.

6. What are the advantages of sending digital messages into space using radio signals? What are the disadvantages?

7. What are the advantages of sending a message into space using spacecraft? What are the disadvantages?

8. Based on your responses to Questions 6, 7, and 8 from Part B, and to Questions 6 and 7 above, what do you feel is the best way of transmitting a message to another civilization? Why?

PART D: COMMUNICATING WITH EXTRATERRESTRIAL LIFE – SHOULD WE DO IT? (OPTIONAL)

Throughout the ages, humans have pondered the question "Are we alone?" We have debated the probable existence of life beyond Earth, searched for evidence of such life, and discussed the ramifications of its discovery. As a group, you will now consider many of the same questions that humans have throughout history. There are no right or wrong responses to many of these questions;, however, it is important to fully explain the reasoning behind all of your answers.

1. Suppose we were to confirm the existence of intelligent life elsewhere in the universe. If we could elect only one spokesperson for our planet:

   a. What country should he or she be from? Explain your reasoning.

   b. What type of qualifications should this spokesperson have (politician, religious leader, athlete, scientist, linguist, artisan, etc.)? Why? What criteria did your group use to come to this decision?

Activity 14

    c. What should the spokesperson say in his or her first communication?

2. A large percentage of Earth's population holds a belief in some form of higher being or power that supercedes the existence of humans. How would the discovery of an intelligent civilization beyond this planet influence these beliefs?

3. We have been unintentionally broadcasting powerful radio signals into space since the 1930s. However, we have intentionally been transmitting information since the launch of the *Pioneer* spacecraft in the 1970s.

    a. What are the positive consequences of sending information about human life on Earth to other civilizations?

    b. What are the negative consequences?

4. Based on your response to Question 3, do you think we should be intentionally trying to contact other intelligent civilizations?

Is It Science?

INDIVIDUAL CHARACTERISTICS BY BIRTH DATE AND HOROSCOPE BIRTH SIGN

Sagittarius Nov 22 – Dec 21	*Leo* Jul 23 – Aug 22	*Aries* Mar 21 – Apr 19
1. LIKES: Optimistic and freedom-loving people; Good-humored jokes; Honesty; Intellectualism DISLIKES: Blindly optimistic people; Carelessness; Irresponsibility; Restlessness	5. LIKES: Speculative ventures Lavish living; Pageantry and grandeur; Children; Drama DISLIKES: Doing things safely; Ordinary, day to day living; Small minded people; Penny pinching; Mean spiritedness	9. LIKES: Action; Coming in first; Challenges; Championing causes; Spontaneity DISLIKES: Waiting around; Admitting failure; No opposition; Tyranny; Other peoples advice
Capricorn Dec 22 – Jan 19	*Virgo* Aug 23 – Sep 22	*Taurus* Apr 20 – May 20
2. LIKES: Reliability; Professionalism; Knowing what you discuss; Firm foundations; Purpose DISLIKES: Wild schemes; Fantasies; Go-nowhere jobs; Ignominy; Ridicule	6. LIKES: Health foods; Lists; Hygiene; Order; Wholesomeness DISLIKES: Hazards to health; Anything sordid; Sloppy workers; Squalor; Being uncertain	10. LIKES: Stability; Being Attracted; Things Natural; Time to Ponder; Comfort and Pleasure DISLIKES: Disruption; Being pushed too hard; Synthetic or "human made" things; Being rushed; Being indoors
Aquarius Jan 20 – Feb 18	*Libra* Sep 23 – Oct 22	*Gemini* May 21 – Jun 21
3. LIKES: Fighting for causes; Dreaming and planning for the Future; Thinking of the past; Good companions; Having fun DISLIKES: Full of air promises; Excessive loneliness; The ordinary; Imitations; Idealistic	7. LIKES: The finer things in life; Sharing; Conviviality; Gentleness DISLIKES: Violence; Injustice; Brutishness; Being a slave to fashion	11. LIKES: Talking; Novelty and the unusual; Variety in life; Multiple projects all going at once; Reading DISLIKES: Feeling tied down; Learning, such as school; Being in a rut; Mental inaction; Being alone
Pisces Feb 19 – Mar 20	*Scorpio* Oct 23 – Nov 21	*Cancer* Jun 22 – Jul 22
4. LIKES: Solitude to dream in; Mystery in all its guises; Anything discarded to stay discarded; The ridiculous; Likes to get 'lost' DISLIKES: The obvious; Being criticized; Feeling all at sea about something; Know-it-alls; Pedantry	8. LIKES: Truth; Hidden Causes; Being involved; Work that is meaningful; Being persuasive DISLIKES: Being given only Surface data; Taken advantage of; Demeaning jobs; Shallow relationships; Flattery and flattering	12. LIKES: Hobbies; Romance; Children; Home and country; Parties DISLIKES: Aggravating situations; Failure; Opposition; Being told what to do; Advice (good or bad)

Adapted from http://www.astrology-online.com